GREAT DISCOVERIES IN SCIENCE

# Carbon Dating

Kaitlyn Duling

Cavendish Square
New York

Published in 2019 by Cavendish Square Publishing, LLC
243 5th Avenue, Suite 136, New York, NY 10016



First Edition



Library of Congress Cataloging-in-Publication Data

Names: Duling, Kaitlyn, author.
Title: Carbon dating / Kaitlyn Duling.
Description: First edition. | New York : Cavendish Square, [2018] | Series: Great discoveries in science | Audience: Grade 9 to 12. | Includes bibliographical references and index.
Identifiers: LCCN 2018014938 (print) | LCCN 2018018438 (ebook) | ISBN 9781502643735 (ebook) | ISBN 9781502643834 (library bound) | ISBN 9781502643957 (pbk.)
Subjects: LCSH: Radiocarbon dating--Juvenile literature. | Geochronometry--Juvenile literature. | Geological time--Juvenile literature.
Classification: LCC CC78.6 (ebook) | LCC CC78.6 .D85 2018 (print) | DDC 551.7/01--dc23
LC record available at https://lccn.loc.gov/2018014938

Editorial Director: David McNamara
Editor: Jodyanne Benson
Copy Editor: Michele Suchomel-Casey
Associate Art Director: Alan Sliwinski
Designer: Christina Shults
Production Coordinator: Karol Szymczuk
Photo Research: J8 Media

The photographs in this book are used by permission and through the courtesy of:
Cover, Didier Descouens, Own work/File: Sahelanthropus tchadensis - TM 266-01-060-1.jpg/Wikimedia Commons/CCA-SA 4.0 International; p. 4 Gavin Shire/USFWS Mountain-Prairie (https://www.flickr.com/people/51986662@N05)/File: Carved ivory tusk stockpile - 2013 USFWS ivory crush.jpg/Wikimedia Commons/CCA-2.0 Generic; p. 10 Michael Maggs, Own work/File: Archaeological dig, Bekonscot.jpg/Wikimedia Commons/CCA-SA 2.5 Generic; p. 13 Microgen/Shutterstock.com; p. 18 Tomekbudujedomek/Moment/Getty Images; p. 21 Stuart Freedman/In Pictures/Getty Images; p. 23 Claudio Divizia/Shutterstock.com; p. 25 Lerner Vadim/Shutterstock.com; p. 28 LBNL/Science Source; p. 30 Don Farrall/Photographer's Choice RF/Getty Images; p. 33 Exeter Paul, Own work/File: Radioactive decay of Carbon-14 SVG.svg/Wikimedia Commons/CCA-SA 3.0 Unported; p. 39 AURORE BELOT/AFP/Getty Images; p. 44 Dmitry Polonskiy/Shutterstock.com; p. 48 Jack Fields/Science Source; p. 51 Galerie Bilderwelt/Getty Images; pp. 52, 70 Bettmann/Getty Images; p. 56 Universal History Archive/UIG/Getty Images; p. 59 MPI/Getty Images; p. 64 Hulton-Deutsch Collection/Corbis/Getty Images; p. 67 Ullstein Bild/Getty Images; p. 73 Bauhaus1000/DigitalVision Vectors/Getty Images; p. 77 Sotheby's Paris, 23 Juni 2011 Lot 30 (http://www.sothebys.com/de/auctions/ecatalogue/2011/tableauxanciens-et-du-xixe-sicle/lot.30.html)/File: Noah's Ark on Mount Ararat by Simon de Myle.jpg/Wikimedia Commons/Public Domain; p. 79 Giuseppe Enrie, 1931/From the Hebrew Wikipedia/File: Shroudofturin.jpg/Wikimedia Commons/Public Domain; p. 82 W. Robert Moore/National Geographic/Getty Images; p. 85 Meunierd/Shutterstock.com; p. 92 Mauro Fermariello/Science Source; p. 98 © NativeStock/North Wind Picture Archives; p. 101 Torben Rick/ Smithsonian Institution; p. 103 Baltimore Heritage (https://www.flickr.com/people/44034762@N04)/File: Animal bone, nail and ceramic artifacts, Patterson Park (16336800175).jpg/Wikimedia Commons/CCA-SA 2.0 Generic; p. 106 Rohitjahnavi, Own work/File: Homo neanderthalensis 2.jpg/Wikimedia Commons/CC0 1.0 Universal Public Domain Dedication.

Printed in the United States of America

# Contents

Carbon dating can be used to learn the age of different types of objects, including these elephant tusks.

# Introduction

One of the most enduring animal-rights issues of the twenty-first century is the buying and selling of elephant tusks. Though the international sale of ivory has been banned since 1990, poachers are still killing elephants, and smugglers continue to bring ivory from those kills into Asia, Europe, and North America. Antique ivory and tusks acquired before 1990 were sellable for a time, but in 2016, a near-total ban on the trade of African elephant ivory was enacted, restricting most legal sales. Since then, illegal sales of ivory have continued at a startling pace. Researchers have been eager to figure out whether the ivory confiscated from illegal sales is from elephants killed recently or decades ago. Then they can get a handle on the current poaching rates, which have been difficult to track, historically.

In order to determine when an elephant was killed, scientists are now able to use a technology developed in the 1940s called radiocarbon dating to analyze the radioactive carbon that remains in the ivory. Then the scientists can figure out how long ago the animal was killed. Once they have that information, those working in the field are much more knowledgeable and better equipped to detect recent poaching and put an end to the practice. In this way, carbon dating has proven, once again, that it is an invaluable resource to scientists, researchers, investigators, and those who are working to combat harmful industry practices. As we will see, it is not only useful to those solving the problems of today, but the method is also widely used to learn about our collective past.

## WHAT IS CARBON DATING?

No, carbon dating does not refer to the romantic lives of carbon. We'll dive deeper as we go along, but basically the term "carbon dating" refers to a scientific process that helps to determine the age of organic matter. Why would we want to do that? Think about the artifacts you see when you walk through a museum. You might visit an exhibit about ancient Egypt that is filled with mummies, sarcophagi, and decorative art. You could look at ancient dinosaur fossils, crumbling paper texts uncovered in Jerusalem, or bowls and pots discovered near a riverbed in the American Midwest. Without processes like radiocarbon dating (also known as carbon dating or carbon-14 dating), we would have little to no understanding of the original time period in which those objects existed. We would have no historical context for archaeological discoveries. The stories would be incomplete, and what we can learn from them would be sharply lessened.

The science behind carbon dating is a little complicated but not impossible to understand. It was first developed in the 1940s by a man named Willard Libby, who won the

Nobel Prize in Chemistry in 1960. Today, carbon dating has become a standard procedure. It is relatively inexpensive, quick to complete, and accessible through laboratories that are scattered across the world. Before carbon dating, we had other scientific processes for determining the age of objects, but now that we know there's a better and more accurate way, there's no turning back! Some have even called the discovery of radiocarbon dating a "revolution" in science.

In addition to profoundly impacting the work that can be done today, such as dating elephants' tusks, it has also had a huge impact on archaeological research. Archaeology deals primarily with human history and prehistory. Archaeologists excavate (or dig) in sites and then analyze the artifacts, monuments, and inscriptions they find there. They might find tools, jewelry, or even entire walls filled with art or words. Using carbon dating technology, these scientists can quickly and accurately determine the ages of the objects they find. The technique doesn't just help us put dates to small objects. It has also allowed researchers to put more accurate dates on important time periods and transitions in history, such as the end of the last ice age or the Bronze Age.

Since it was developed in the late 1940s, carbon dating has become the standard method for dating artifacts, often helping to add context and timelines to our previous understandings of world history and ancient history. However, this method does have its drawbacks and limits. For instance, it can only be used if the object being dated contains carbon-bearing ingredients. Archaeologists have had a hard time dating some ancient cave paintings, located in Spain.

## LIMITS and POSSIBILITIES

The images on the Spanish cave walls are abstract, including lines, dots, and the outline of a human hand. It was long believed that these paintings were made by early modern humans. Radiocarbon dating could not be used to date the paintings themselves because red ochre does not contain carbon-bearing ingredients.

So scientists relied on the carbon dating of shells in the same layer of rock. They dated to about forty-five thousand to fifty thousand years. Unfortunately, carbon dating becomes very unreliable beyond forty thousand years. That means the shells could be much older. In this case, carbon dating was not the most accurate or reliable way to date the paintings.

Scientists turned to uranium dating, a different method, to test the cave art. The results show that Neanderthals may have been responsible for the art, which is a huge breakthrough in evolutionary study. Sometimes carbon dating is just one step on the path to other methods that will give us accurate and helpful answers.

So, we know that carbon dating can help us put dates and context to the past. How does that help us now, in the twenty-first century? Dating technology can have significant implications for our present-day behaviors and decisions. For example, if we are able to better understand when, where, and how ice ages have occurred throughout history and how long they lasted, we may be better able to contextualize our current climate situation. Those who believe that the planet is warming at an unnatural rate are often at odds with those who see global warming as a natural phenomenon.

Carbon dating can give us a view into the past, giving us a picture of those events that may or may not be cyclical or naturally occurring. Once we have a handle on that information, we can better plan for natural changes in weather while also making crucial decisions about the ways in which we contribute, as humans, to climate change.

When it comes to planning for our future, it is crucial that we first understand our past. Carbon dating is one of the best tools we have for making that happen. In the next chapter, we will explore the problems that led to the discovery of carbon dating and the ways carbon dating was used to address those problems.

Sometimes workers at archaeological dig sites use carbon dating and other techniques to learn more about places and the people who lived there long ago.

CHAPTER 1

# The Problem of Dating Our Earth and Its Organisms

Think back into your memory. Can you describe what life was like ten years ago? Now, can you imagine what life was like one hundred years ago? How about four thousand years? Fifteen thousand? What about millions of years ago? On a day-to-day basis, we focus on the here and now. We text our friends, do homework, eat dinner, and sleep. It is easy to contemplate the near past, such as our own childhoods. Even time periods, such as the American counterculture of the 1960s or the pain of the Great Depression in the 1930s, are fairly simple to access from the history we've been taught. But when you stretch your mind back thousands, millions, and even billions of years into the past, it becomes more and more difficult to comprehend those long stretches of time, what they looked like, and what they mean. But scientists, archaeologists, biologists, astronomers, and others working in similar fields have always been on the hunt for more information about the times that came "before"—before photographs, before writing, before modern human records.

The records used by workers in these fields don't look like books or paintings. Historical records only reach back about five thousand years, around the time of the ancient Egyptians. Instead, they are left with fossils of bugs and bones; pieces of structures, such as houses; or even the great pyramids. In order to glean answers from these types of records, researchers have had to employ a variety of methods. Instead of simply reading about history or watching a film about history, they have to do the history, discovering the answers for themselves.

## RELATIVE DATING

Prior to the mid-twentieth century, the only method for identifying the age of an ancient object was relative dating. With relative dating, scientists can determine the relative order of events and then enter that object into the context. This does *not* give us the absolute age of the object. Relative dating has been used most heavily by archaeologists and geologists because rock, fossils, and deposits can help determine relative ages. It is important to remember that relative dating does not tell us when an event occurred or when an object was formed. Instead, it tells us the order in which a series of events occurred. How does relative dating accomplish this?

When an archaeologist uncovers an object, such as a coin or a known piece of artwork, it is fairly simple to figure out its age. The coin may have a date on one of its sides. The artwork may have come from a specific, known period and location. However, when an excavator uncovers a fossil

This archaeologist is uncovering a fossil buried deep within one of Earth's many layers.

or another object without its own date or dated context, researchers have to do a little investigating and, ultimately, guess the age of the item. When identifying the age of an object with relative dating, scientists are determining that the object is older or younger than something else. They do this using layers and layers (and more layers) of rock. These layers are called strata.

Usually, layers of rock are laid down in order, from bottom to top, oldest to youngest. It makes sense if you think about the dirt in your backyard. Whatever is on top would be the newest, blown over from elsewhere. The soil at the bottom is the oldest. It is the same with rocks. The principle of "oldest to youngest, bottom to top" is called superposition. Scientists also use the principle of horizontality, which tells us that most rock is laid down flat in horizontal layers. So, if you cut out a uniform piece of rock, perhaps a foot long by a foot wide, but extending many feet downward, you would be able to see the different layers that occurred over time. Each layer would probably have its own color, texture, thickness, and chemical makeup—almost like the layers of a cake.

Scientists don't just look at these strata in order to admire the amazing colors. They use them, in combination with fossils, to get an understanding of relative time. A fossil is the impression or the remains of an organism—perhaps a bug, animal, or plant—that is molded or cast into rock. When a bug dies on the ground, its body can be pushed down hard into the rocks there, pressing an imprint of it into land that will harden and stay over time. This process takes thousands to millions of years. When we call something a fossil, we usually refer to things that died over ten thousand years ago. However, you can make something that looks like a fossil by simply pressing a leaf into some clay. A fossil is just a very old version of your leaf print.

Geologists have worked with fossils to figure out the order in which living things (also called organisms) appeared and disappeared. Even when they are found a long way apart, fossils that are discovered in the same layer (strata) of rock are considered of the same time period. This geological science wasn't formed until the 1800s—only two hundred years ago! Relative dating generally works well in order to place events on an historical timeline even though it doesn't produce an absolute age. From relative dating, scientists are able to determine when a major event happened relative to one another, such as the dinosaurs and the ice ages, or which creatures did or did not live during the same time periods. For a long time, this was the only method we had for dating objects that were found in the ground. There was no process that could tell us when, exactly, an object was created or was living.

## DENDROCHRONOLOGY

You may have heard that you can tell the age of a tree by counting the rings inside its trunk. But how exactly does that work? Dating through tree rings is actually a scientific method that we refer to as dendrochronology. Tree rings can be dated to the exact years in which they were formed, giving scientists insight into major weather events, climate changes, atmospheric conditions, and more. Some trees are extremely old, lending scientists an even longer timeline from which to learn.

# Unique Dating Methods

Over the years, scientists have developed some surprising techniques for dating. One of these is palynology—a method that utilizes bee pollen. It works like this: Seed-bearing plants release huge numbers of pollen grains each year.

We now know when seed-bearing plants release huge numbers of pollen grains because that is a heavy allergy time for many people. A "rain" of pollen, though you cannot always see it, falls over environments like fields and lake beds. Sometimes, if left undisturbed, that pollen can even become fossilized. Then, scientists can figure out which species of pollen were deposited at different times based on where the pollen resides within rock layers, effectively creating a "pollen calendar" of sorts.

The period of time in which a plant species is more abundant than other species can be noted by using palynology techniques. For example, a field may have been full of wildflowers for many years, but a scientist can detect from the fossilized pollen when the field turned into one planted with grain or grass. The period of time is called a pollen zone. Of course, palynology is a form of relative dating that cannot accurately predict absolute dates.

Another unique form of relative dating is the use of ice core samples. Since the early 1900s, scientists have drilled into ice sheets and glaciers to remove core samples. Ice forms from the gradual, year-by-year buildup of snow. Just like rock or untouched soil, the oldest layers of an ice core can be found at the bottom.

Ice cores can contain ice that is hundreds of thousands of years old, reaching miles into the earth. Ice cores are primarily used to detect changes in climate. Researchers examine the density and size of bubbles as well as the visual layers to detect changes in everything from atmospheric gases to seasonal temperature fluctuations.

The date markers on this cross section of a tree help build a timeline of the tree's life.

A tree grows taller, but it also grows in diameter. It gets wider. Botanists have been talking about this since it was first suggested in ancient Greece, but it wasn't until the 1800s that the science of tree rings really took off. The rings you see in a cross section of a tree are the result of growth in a layer of cells very near the bark. One ring usually represents one year in the tree's growth. However, extreme conditions can result in multiple rings forming in a single year or even missing rings. For the most part though, tree rings are a fairly accurate source of absolute dating, especially because dendrochronologists have been able to compare trees across large areas, building reference points for changes in climate and weather. Their use isn't simply to date the trees themselves. It also helps scientists figure out the ages of other structures, such as homes and buildings uncovered on archaeological digs.

When wooden ruins are found, trees can help researchers build a chronology that puts a building into a relative timeframe. They do this by extracting the cores of both living and dead trees in the area and dating them carefully. Then, a wooden beam or another portion of the wooden ruin can be cross-dated against the tree rings. This type of dating has also been used with the wooden panels used to make paintings between the fourteenth and seventeenth centuries.

One outgrowth of tree ring study has been acanthochronology, which uses cactus spines to put growth into a time context. Like trees, which tend to grow continually, cactus spines keep growing, shifting, and growing. Generally, the oldest spines are at the bottom and the new ones end up at the top. By studying cacti, researchers can learn about the climate and environment. This field of study is relatively new but builds on centuries of tree ring experience and study.

## PALEOGRAPHY and EPIGRAPHY

Another method that has been used over the centuries for the purpose of dating is the study of ancient writing. While it cannot be used to accurately pinpoint exact dates, it has been invaluable to historians, many of whom would someday utilize radiocarbon dating to place their work even more accurately on the world timeline.

Paleography (the study of handwriting) doesn't simply put different writing styles into a diagram of dates. It is a careful study that rests on language, historical writing styles, grammar, alphabets, and abbreviations. Handwriting and language are constantly changing throughout time and place, so it is necessary for researchers to be well versed in the particularities of different time periods and locales. A good paleographer can provide a wealth of information about the time and place in which a document may have been written. This study has been utilized in order to date texts from all over the world, some of which are related to religious texts, governmental papers, and other texts that give us insight into how and why different cultures functioned the way they did.

Greek, Latin, and other ancient languages that were usually written on papyrus, parchment, or paper are often the subjects of this type of dating work. The paleographer assigns date and place of origin to the document after reading it and considering it against acquaintance texts, taking into account the numerous differences in alphabet, writing styles, abbreviations, and more. An expert must also be aware of changes in the calendaring system that occurred in different points throughout history, as calendars have been reformed multiple times. Moreover, the dates included within a text may not be an accurate reflection of how we understand dates today. Above all, one who is utilizing this method must have a treasure trove of information at hand on which to form his or her opinion

Researchers like this one in Pondicherry, India, study ancient religious manuscripts. These Hindu texts were written on palm leaves.

about the date of the document. An inaccurately dated document can reverberate throughout the world, especially if it holds emotional, political, or spiritual significance for the lives of many, such as ancient texts related to the Bible, Quran, or other religious book.

Epigraphy is a closely related study, but it is focused on drawing conclusions and dates from inscriptions, also called epigraphs, on buildings, statues, coins, and more. These could be engraved into rock, painted on ceramic, embossed in metal, or even impressed into wax. While paleography deals with handwriting usually written in ink, epigraphy covers many different types of writing—stamped, carved, scratched, etc. Just like archaeologists, geologists, and others who work with the dating of objects, epigraphers must develop strong knowledge bases about time periods, locations, and cultures in order to accurately give a date to an inscription. Deciding the date of a piece of rock or clay tablet based solely on an inscription can take a long time, and it is not always accurate.

There are three different types of inscriptions that are often studied—monumental, archival, and incidental. A monumental inscription is usually made out of stone or metal, as it was meant to survive and serve as a visual piece for many people to see and read. Monumental inscriptions might appear on tombs, walls, etc. Archival inscriptions were used within early societies in order to keep records. Much like we keep governmental records and even personal histories today, early cultures wrote on clay tablets and other materials in order to archive the events and ideas present within their society. Incidental inscriptions were not meant to be preserved. Though much

The Rosetta stone, which is inscribed with ancient Egyptian and Greek languages, helped researchers understand Egyptian hieroglyphics.

ancient writing was done with care and intention, some was just scribbled out onto a wall or piece of papyrus, such as a note or even an ancient form of graffiti. These casual inscriptions are no less interesting or of less importance to epigraph professionals. They can tell us quite a bit about a society, culture, place, and time. These types of writings, especially of the archival sort, can be particularly helpful for chronological ordering, which helps when it comes to putting other objects within a time period.

Inscriptions, like handwritten documents, are especially important objects when it comes to developing a chronology (timeline) of events in time. An inscription can combine not only an object that can be relatively dated using strata, but it has content that lends historical clues about what was happening at the time. An epitaph might tell us about the death of a political or religious leader. Some of the earliest deciphered inscriptions are the edicts of Ashoka, which were inscribed on pillars, boulders, and grave walls between 269 and 232 BCE. These are some of the very earliest written materials on the Indian subcontinent and are also the first tangible evidence of Buddhism. By studying these inscriptions, historians understand how and where languages developed over time and develop a general chronology for the area.

## NUMISMATICS

One very particular form of studying dates and chronology is through numismatics, which includes the collection of money—coins, paper money, tokens, and even shells, beans, stones, and gems, which have been used as currency throughout history. Though this field of study is often populated with coin collectors and those most interested in post-seventeenth-century currency that is printed with machines, numismatics can also be helpful when it comes to dating and chronology work. When archaeologists are digging in a particular spot and find a coin, that coin may have a date on it or it might be associated with a particular government. These are easy ways to place the coin and thus other objects uncovered in the same strata within a timeframe.

## TOWARD A NEW WAY

As you can see, the period of time before radiocarbon dating was a very long one, fraught with inaccuracies in dating methods and an excess of information that was only relative, rather than absolute. Historians and archaeologists relied heavily on the idea that old things are buried under younger things. Stratigraphy tells us this and gives us a framework for making a loose chronological timeline, but it does not accurately date materials by year, decade, or even century.

As late as the 1800s, Rasmus Nyerup (a Danish historian) wrote that early human history was "wrapped in a thick fog." He went on to say, "We know that it is older than Christendom, but whether by a couple of years or a couple of centuries or even by more than a millennium, we can do no more than guess." It wouldn't be until the 1940s that science

Portions of the Dead Sea Scrolls have been placed on display for the public to see.

would start to move toward a more accurate way to quantify archaeological time.

One discovery that happened at the very cusp between the creation of radiocarbon dating and the reliance on old forms of dating was the discovery of the Dead Sea Scrolls. Often referred to as one of the greatest manuscript discoveries of all time, these texts are made of animal skins, papyrus, and copper and were transcribed with a carbon-based ink.

The remains of approximately 850 documents, the Dead Sea Scrolls were slowly discovered between the years 1947 and 1956. They were discovered in a string of eleven caves on the shore of the Dead Sea, about 13 miles (20.9 kilometers) east of Jerusalem. These texts include biblical manuscripts from the books of the Old Testament, commentaries, rules of war, psalms, sermons, hymnals, and community regulations.

They are written in Hebrew, Aramaic, and Greek. It is probable that the scrolls were the library of a Jewish sect living in the area. When they were discovered, everyone wanted to know: How old were the scrolls? Were they real? Were they forged and planted in the caves recently, or were they truly from ancient times?

The first date analysis that was performed on the Dead Sea Scrolls was paleography, the study of handwriting. Multiple scholars in the field examined the scrolls. By examining the size, style, and variability of the text, paleographers were able to determine that the texts were probably written between 225 BCE and 50 CE. A range of 275 years may not be as accurate as possible, but it gave researchers a window in which to work and a shocking one at that.

Until the discovery of the scrolls, the oldest Hebrew-language manuscripts of the Bible dated to the tenth century CE. When the scrolls were given their date, they became the oldest by over one thousand years, prompting scholars to reconsider countless beliefs and theories about the Old Testament.

It would be a few years before the radiocarbon revolution was ready to take on the Dead Sea Scrolls and provide a more accurate date. Until then, scholars and all those watching and wondering about their origins would have to make do with paleography and other relative analyses. Soon enough, though, Willard Libby's discovery would add another possible method for dating the scrolls, while at the same time changing the world of archaeology forever.

Built in 1930, this initial working model of Ernest Lawrence's cyclotron would lead to great advancements in carbon dating technology.

CHAPTER 2

# The Science of Carbon Dating

For most of the earth's history, accurate dating of ancient objects was not an option. We did not yet have the technology or scientific understanding that would allow us to learn about an object's or material's life using chemicals and mathematics. However, in the mid-twentieth century, physicists and other scientists began to make the breakthroughs that would lead to the use of radiocarbon dating methods in order to date the Dead Sea Scrolls, among countless other objects. In order to fully understand and appreciate Willard Libby's great discovery, though, we must first begin with the early research that was the foundation for carbon dating experiments, and we must review the basic science of the method itself.

## CALIFORNIA DREAMIN'

In 1931, the Lawrence Berkeley National Laboratory (LBNL or LBL) was built at the University of California, Berkeley. At first, it was referred to as the Radiation Laboratory and

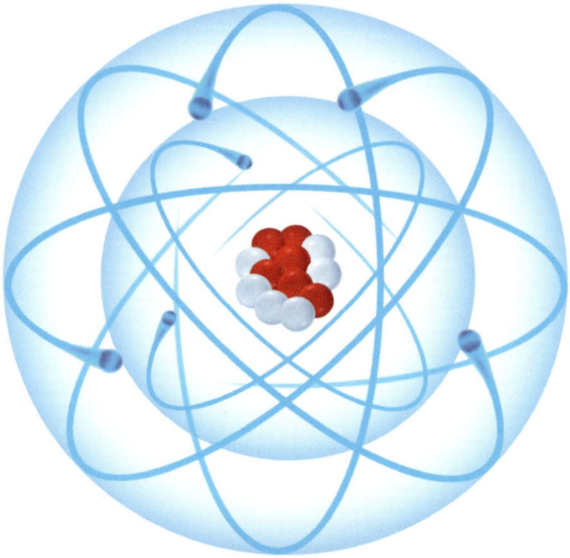

This is a carbon atom. Inside, you can see its six protons, six neutrons, and six electrons.

centered its physics research around the cyclotron, a particle accelerator machine. The laboratory would eventually become the home of several of the key players involved in the Manhattan Project, the plan to construct and deploy the first atomic bombs. By 1939, Martin Kamen and Samuel Ruben, two young scientists, were working together at the lab, conducting experiments on elements to see if they could be useful for biomedical research.

Ruben and Kamen began to use the laboratory's cyclotron accelerator to do experiments on carbon. Carbon is an element on the periodic table. There are about 118 known elements, which are the simplest form of substances we have in our world. Everything—all matter—is made up of these elements. When we want to see a diagram with all of the elements laid out next to each other, we look at the periodic table.

Elements are made up of atoms, which are the basic building blocks of matter. They are so tiny, it is difficult to imagine them. You may remember that an atom has a middle that we call the nucleus. Inside the nucleus there are protons and neutrons. Moving in orbit outside the nucleus are electrons. We refer to protons, neutrons, and electrons as subatomic particles. We distinguish the different elements by their atomic numbers. This refers to the number of protons the element has in the nuclei of its atoms.

Remember carbon? Carbon is a nonmetallic element found widely, though not plentifully, in nature. It has six protons, six neutrons, and six electrons, which means that its atomic number is six. Atoms with six protons are forms of the element carbon. If an atom has seven protons, it is a form of the element nitrogen. One fun part about elements is that they aren't as simple as they seem. They can come in different forms. When the atom of an element has the same number of protons but a different number of neutrons, it is called an isotope. Carbon itself is not an isotope because it has six protons and six neutrons. Because of this, we can also call it carbon-12. But there are hundreds of atoms with differing numbers of neutrons and thus hundreds of isotopes; each element has multiple isotope forms. Carbon, of course, has several isotopes. One of those isotopes is called carbon-13. This is a natural, stable isotope. It will never change into another isotope or element on its own. Inside its nucleus, it contains six protons (just like carbon) and seven neutrons.

What Martin Kamen and Samuel Ruben discovered, way back in 1939, was a special isotope of carbon. On February 27, 1940, Ruben and Kamen discovered the isotope using the cyclotron accelerator. The substance is known as carbon-14, $^{14}C$, or radiocarbon. It is often called radiocarbon because it is a radioactive isotope. What makes something radioactive? When an atom has excess nuclear energy, it can become unstable, able to transform or decay into another type of atom. That excess energy can be emitted as radiation, released as an electron, or used to create a new particle. The cyclotron at Berkeley helped to produce carbon-14, which is radioactive. Radiocarbon's nucleus contains six protons and eight neutrons. That's where the excess nuclear energy comes from.

## HALF-LIFE

Now you are well on your way to becoming a nuclear physicist, just like Kamen and Ruben. When they synthesized carbon-14, they were hoping to study the isotope's half-life. Put simply, a half-life is the time it takes for a quantity to reduce to half its starting value, on average. To find the half-life of a glass of milk, you would see how much time it usually takes to drink (or pour down the drain) exactly half the milk in the glass. This principle was discovered in 1907 by a man named Ernest Rutherford.

Here's another way to think of half-lives: When you have 100 percent of something, a half-life has not elapsed. When you have 50 percent remaining, one half-life has elapsed. When 25 percent is left, two half-lives have elapsed. We know that two have elapsed because the original 100 percent

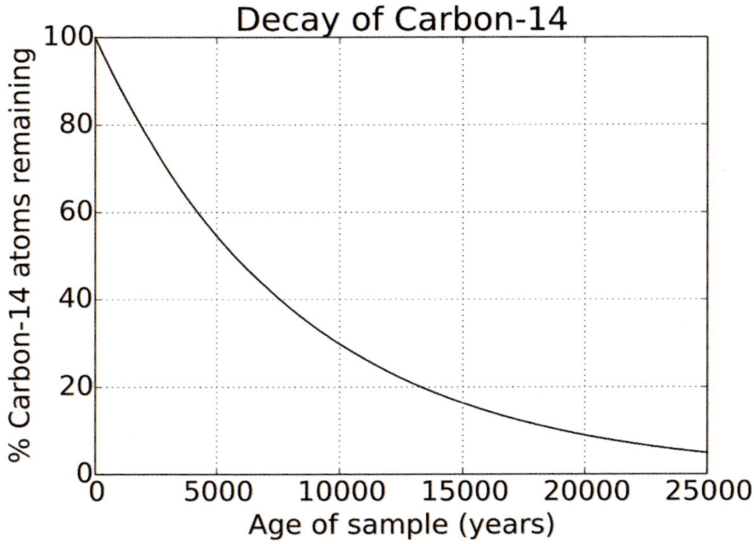


This table shows the decay rate of carbon-14. As the years go by, fewer atoms remain.

has been cut in half—twice. You might think that a material or element could only have a few half-lives, but things can keep dividing, on and on, without ever disappearing. In nuclear physics, we use half-life terminology to discuss how long it takes unstable atoms to undergo nuclear decay. We can also call the decay process nuclear decay or radioactivity. You've probably heard the term "radioactivity" before. On television, characters sometimes encounter radioactive materials. Now you know that this means that the nuclei have become unstable and begun to transform or "decay," emitting radiation. Different atoms will have different decay rates, which we describe by using the number of half-lives it has. Some radioactive atoms decay almost instantly.

Others have more half-lives than we can count, stretching longer than the age of the known universe.

It's key to note that half-life is more about probability and less about quantity. If we have one atom, and it becomes radioactive and has a half-life of one second, what does that mean? It does not mean that half of the atom will disappear in one second, finished with its decay process, and the next half will be gone in the next second. What it does mean is that if we wait for one half-life of one radioactive element, then there is a 50 percent chance it will decay. The science is not exact. After one second, there would be approximately one-half of the atom left, but not always and not precisely.

## The HALF-LIFE of CARBON-14

Remember, carbon-14 only occurs naturally in very small, trace amounts. As a radioactive material, it experiences radioactive decay. The type it goes through is called beta decay. During this process, the carbon-14 atom's nucleus emits an electron and an electron antineutrino, a particle produced during nuclear reactions. This electron antineutrino does not live inside the atom all the time—just during this reaction. When this emission happens in a carbon-14 atom, one of the neutrons inside the atom decays into a proton, and the carbon-14 atom transforms (or decays) into the isotope nitrogen-14, which is stable and nonradioactive. So, when radioactive decay occurs for carbon-14, it turns into nitrogen-14 and becomes stable. It is also true that this decay occurs in a fairly predictable fashion, as it does with most unstable atoms. While there are exceptions, the half-life for carbon-14 is about 5,730 years, with a margin of error of about 40 years. If we took a sample of carbon-14 and waited 5,730 years, half of the atoms would have decayed into nitrogen. In another 5,730 years, half of the remaining atoms would have decayed into nitrogen, and so on.

Prior to Kamen and Ruben's experiments at the University of California, Berkeley, it was thought that the half-life of carbon-14 was much shorter. Using the cyclotron, they discovered the number 5,730. Without their work, the scientific discovery of the radiocarbon dating process could not have occurred, at least not when and where it did. They were the first to discover and create the carbon-14 isotope. Though they would go on to be accomplished scientists, contributing to the Manhattan Project, inventing things, and even founding Duracell, the two made a notable contribution to science and history with their discovery of carbon-14.

## SERGE KORFF'S PREDICTION

While Kamen and Ruben were necessary players in the years leading up to the discovery of radiocarbon dating, they weren't the only ones. Physicist Serge Korff worked at New York University, in New York City. In 1939, while studying at the Franklin Institute in Philadelphia, he made a notable discovery that put in motion the idea of naturally occurring radiocarbon. Korff discovered that neutrons were produced when the atmosphere was bombarded by cosmic rays. The upper atmosphere is full of nitrogen-14, which is what carbon-14 decays into. Korff predicted that when the neutrons reacted with nitrogen-13, carbon-14 would be produced. Before this prediction, scientists had guessed that carbon-14 would be created by the interaction of a stable isotope of hydrogen, hydrogen-2, interacting with carbon-13. Korff's idea was new and interesting, but it wouldn't be fully utilized until 1945, when Willard Libby began his own study of radiocarbon dating.

What Korff's idea did bring to light is that $^{14}C$ does occur naturally; it is not only formed through the processes like those Ruben and Kamen used to produce it in their experiments. In order to understand how it is produced in nature, first we must know about something a little mysterious, invisible, and constant—cosmic rays.

Cosmic rays are high-energy atom fragments (radiation) that pummel Earth from outside the solar system—some from distant galaxies. When they hit the top of the planet's atmosphere, particles rain down. Cosmic rays, which were first discovered in 1912, move faster than the speed of light. They can come from our sun, but much of their origin remains unknown. Sometimes they cause problems for satellites and other technology that we put into the atmosphere. For this reason, scientists are constantly on the hunt to learn more about cosmic rays. Some believe that they are emitted from star explosions.

Wherever they come from, these rays are made up of protons, electrons, and/or atomic nuclei, all of which collide with atoms in the upper atmosphere, very rarely traveling to the ground. However, secondary particles do reach us here on Earth, though they don't necessarily seem to hurt us. Their effect on humans is very poorly known.

We do know, though, that when they enter Earth's upper atmosphere, cosmic rays interact with nitrogen-14 atoms, creating radiocarbon. The isotope is then oxidized and absorbed by all living things, including animals and plants. It is incorporated into plants through photosynthesis, in which plants absorb natural energy, and then animals (including humans) eat the plants. It also enters the carbon pool in oceans, lakes, and rivers, where marine life and shells take in the $^{14}C$ as well. When a plant or animal that contains $^{14}C$ dies, it stops absorbing the isotope because it stops exchanging $CO_2$ with the atmosphere through breathing. Instead, it immediately begins to decay, beginning the 5,730-year countdown. This process of human, plant, and animal intake of carbon-14 is a natural one, but humans have exacerbated it during certain historic periods. In the 1950s and 1960s, excessive testing of nuclear weapons put too many neutrons into the atmosphere, which caused $^{14}C$ to be created. This process went on for years, nearly doubling the ratio of carbon-14 to carbon in the atmosphere, which then added extra $^{14}C$ to the ocean and living things as well.

Korff's work on cosmic rays was necessary to the eventual development of radiocarbon dating. His scholarship successfully showed that carbon-14 occurs naturally in Earth's atmosphere. Without this idea, Libby most likely would not have considered carbon dating because that process assumes that carbon-14 occurs naturally and exists inside organisms.

## MEASUREMENT

In order to figure out the radiocarbon "age" of a given sample, one must first figure out how much carbon-14 the sample contains. There are two ways to do this: the radiometric method and the AMS (accelerator mass spectrometry) method. For many years, the radiometric method was the only option. AMS dating is a relatively new option and arguably the most popular today. For this reason, we will first investigate the radiometric method in order to get a lay of the land.

To date something using the radiometric method, one uses either a gas proportional counter or a scintillation counter. The proportional counter device measures particles of ionizing radiation and the energy they contain. It produces a gas output that is proportional to the radiation energy, which is how one can measure the radiation energy. A scintillation counter, on the other hand, measures particles of ionizing radiation by detecting light pulses. Photons are generated in response to radiation, which are then converted into an electrical signal. The best part about these radiometric dating methods? They are cheap! Today, it costs about $300 per sample, though it takes about a month to achieve results and requires at least 100 grams (3.5 ounces) of the carbon sample in order to detect carbon-14 through the decaying particles. It works best with newer samples that have high beta activity.

The newest and arguably best way to measure the amount of $^{14}C$ within a sample is through the use of an accelerator mass spectrometer. The method itself is referred to as AMS. It differs from radiometric dating in this way: a scintillation counter or proportional counter detects beta

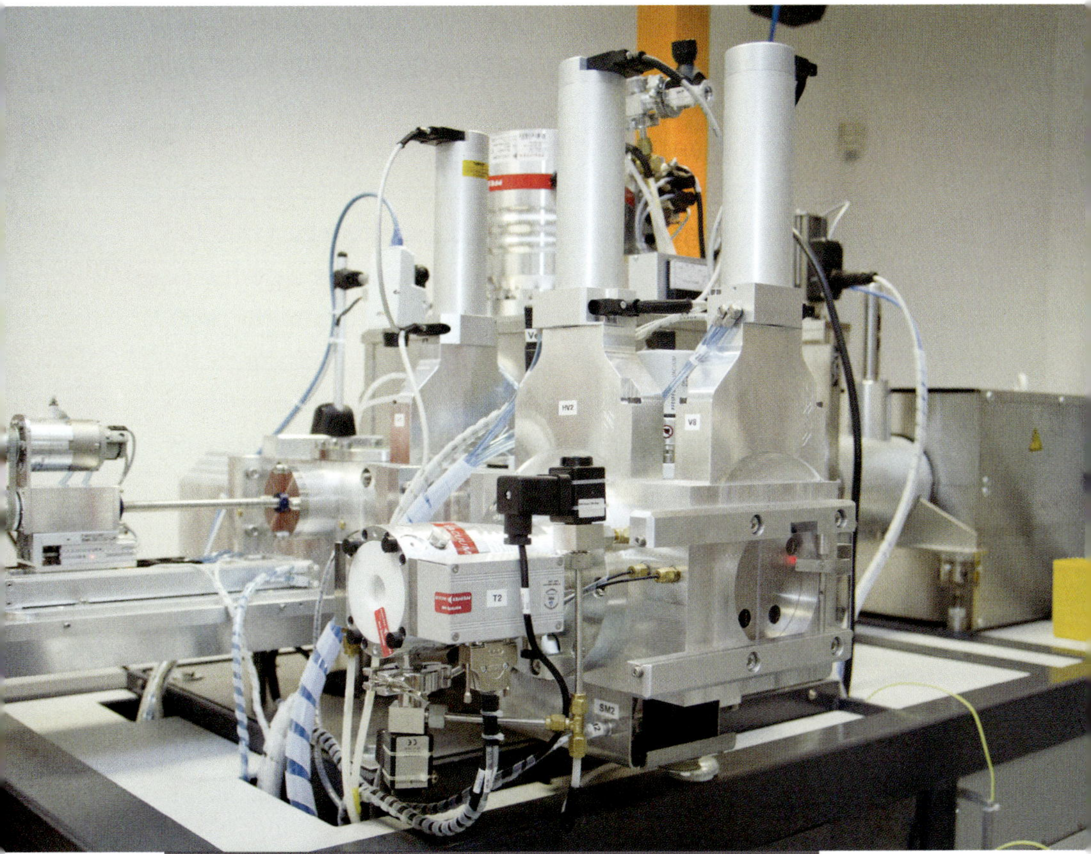

An accelerator mass spectrometer (AMS) like this one can be used to figure out how much carbon-14 exists inside of a given sample.

particles from the decay that is occurring. Accelerator mass spectrometers count the number of carbon-14 atoms that are present in the sample. They are extremely expensive, costing millions of dollars to construct and upkeep, and only a few facilities have them. Universities and companies send their samples to these labs, which then process the samples for a fee.

Essentially it isolates the rare isotope ($^{14}C$) from the mass ($^{12}C$). Oftentimes the machine can also separate nitrogen-14 from carbon-14. Accelerators were first used to detect radiocarbon in the 1970s, and the method grew in popularity throughout the 1980s and 1990s. Today the machines are used for many different biomedical research projects and continue to be used to detect radiocarbon. Besides being expensive, the AMS method uses a very small sample size, which makes control of contaminants a struggle. It costs closer to $600 to complete a sample using AMS technology.

## DETERMINING AGE

As we learned above, plants and animals are exchanging carbon with their surroundings throughout their entire lives, necessarily taking in carbon-14 in addition to carbon-12. The carbon-14 an organism contains has the same proportion of $^{14}C$ as the atmosphere has. At death, the organism no longer takes in $^{14}C$, but the amount contained inside of it will continue to decay and diminish. The $^{14}C$ to $^{12}C$ ratio will decrease as time elapses. We already know the rate of carbon-14 decay: approximately 5,730 years. So, the proportion of carbon-14 is then used to figure out how much time has passed since the organism stopped exchanging carbon. The older a sample is, the less $^{14}C$ will be left within it. This means that radiocarbon dating, at its most simple, is a measure of the level of radiocarbon within a sample, placed in context and noting what we know about the half-life of $^{14}C$. Using mathematical formulas, a researcher can determine the approximate age of a sample.

Of course, in order to be dated using radiocarbon dating methods, a sample must contain carbon-14 and have once been part of a living organism because that is how it would have obtained the radiocarbon. Here are a few things that can (and have) been dated using AMS or radiometric dating: mud, soil, water, wood, shells, seeds, fabric, paper, bones, hair, leather, and even blood residue. Other objects, such as metal, stone, and ceramic pottery are not candidates for radiocarbon dating.

## ERRORS and DRAWBACKS

When attempting to draw conclusions about age through radiocarbon dating, scientists always need to consider the potential issues that come with the method. It is certainly not a perfect science. Oftentimes, there are variations in the $^{14}C$ to $^{12}C$ ratio in the atmosphere itself. This can occur in a given place, but it has also happened historically, changing the levels over time and making the measurements inaccurate from the start. The $^{14}C/^{12}C$ ratio has not remained static over the last few thousand years. On the contrary, it has changed due to many factors, including nuclear testing, the burning of coal and oil, and more. In order to account for these changes, researchers have been able to construct something called a calibration curve. This curve shows

## Serge Alexander Korff

When Willard Libby won the Nobel Prize in Chemistry for his discovery of radiocarbon dating, he took the time in his speech to remind the audience of Serge Korff, whose contributions were invaluable to the science of radiocarbon and cosmic rays. Korff was an educator as well as a physicist, writer, and researcher. In fact, some have called him an ambassador of science, especially due to his extensive travel life.

Born in Finland, Korff immigrated to the United States with his father, who was a baron and belonged to a prominent Russian family. Eventually, Korff ended up at Princeton University, in New Jersey, where he obtained his bachelor of arts, master of arts, and doctorate. As a researcher in California at the Mount Wilson Observatory and the California Institute of Technology, he focused on the observation and mapping of cosmic rays, which are particles from outer space.

He used balloons to study the upper atmosphere, flying them over Guam, Greenland, New Jersey, Connecticut, and many other locations, where they were often reported as UFOs (unidentified flying objects). His students lovingly referred to themselves as Korff's "balloonatics."

He also traveled to mountaintop observatories and rode in high-flying airplanes in order to observe the upper reaches of Earth's atmosphere. No point was too high, or even high enough.

In the 1940s, he began work at New York University, becoming president of the New York Academy of Sciences in 1972. As a researcher, he authored over 175 scientific and technical papers, which wasn't too surprising considering that Vladimir Nabokov, the prolific Russian writer, was his cousin. Even after he retired, he continued to take on leadership roles in various scientific organizations. In 1989, Korff died in his Manhattan home, at the age of eighty-three.

The planet's oceans hold huge reservoirs of carbon.

us the actual $^{14}C/^{12}C$ ratio that was in the atmosphere in a certain year.

The data for the curve was collected, beginning in the 1960s with a man named Hans Suess, from tree rings. Remember those? Each ring in a tree preserves the ratio of $^{14}C$ to $^{12}C$ that was in the atmosphere during any given year.

This data is accurate and naturally occurring, so scientists were able to accumulate the numbers relatively quickly. In fact, we now have tree-ring data for about fourteen thousand years, which corrects for fourteen thousand years of atmospheric variation in the radiocarbon dating process.

Further complicating the method is the exchange between atmospheric and oceanic radiocarbon. The ocean holds a vast reservoir of carbon, both on its surface and within its depths. Radioactive decay occurs in the deep ocean, and then the old carbon mixes with the new carbon in an action called upwelling, which varies in amount in different oceans around the world. The deep ocean and surface ocean have very different apparent ages, and organisms such as shells, coral, and fossils collected from the ocean have variable ages. This is called the reservoir effect.

Scientists consistently study corals and other oceanic organisms in order to accurately measure the effect. When taken together, the "calibration" of radiocarbon using tree rings, leaves, corals, and other organisms means that there is a recalibration curve that dates back fifty thousand years. So, anything dated within the last fifty thousand years can be more accurately assessed with the help of the calibration curve, which will add or subtract years based on the years in which it was living.

Some objects are simply (and unfortunately) not great candidates for radiocarbon dating. For instance, a woolly mammoth leg bone found on Gabriola Island in Canada was going to be carbon dated, and scientists were optimistic. Unfortunately, the bone did not contain enough collagen, which is where the carbon isotope would have been located. Bone is mostly made up of calcium phosphate, so it did not include enough carbon for even a small sample. Researchers can only guess at its age.

## A TRICKY SCIENCE

Radiocarbon dating is not an exact science, and its discovery wasn't simple. It took many scientists over many years, learning about the nature of carbon to finally arrive at the conclusion that organisms could be dated using carbon-14. Of course, the process of radiocarbon dating doesn't give a clean or exact date.

Changes in calendars, reservoirs in oceans, and contamination in samples can all cause issues. Cost can be a barrier. And the ratios of $^{14}C$ to $^{12}C$ in the atmosphere and oceans have changed so much over time that scientists were forced to create an entire calibration chart in order to correct the data. We will learn about other limits to the method in the next few chapters.

When you're thinking about carbon dating, try to remember that it involves more than just taking a sample of a substance and running it through a complicated machine.

It is a chemical process that utilizes what we know about the nature of radiocarbon.

The age "test" occurs at the atomic level, where we can measure the half-life of carbon and then work backwards to determine the real-life age of a sample. Why would anyone go through the expense, time, and effort to engage in such a complex process?

Radiocarbon dating is an extremely valuable tool that has been used to date some of the world's most important ancient artifacts. It has also given us clues to how we should compose ourselves and our lives in the twenty-first century. This microscopic process has huge, life-size implications, and as scientists continue to learn more and perfect the method, it can only become more useful and accurate, providing an enduring method for dating that researchers continue to utilize year after year.

Dr. Willard Libby helped develop the process of radiocarbon dating.

CHAPTER 3

# The Major Players in Carbon Dating

The discovery of radiocarbon dating was a major scientific event that shook up the archaeological community and forever changed the ways in which we are able to conceive of an object's history. To many, Willard Libby is the star of the story, as he was the first to figure out how the process would actually work in real-life applications. However, scientific discoveries are quite often more like structures made of building blocks. One breakthrough builds on another, which builds on another, and so forth. Libby would never have been able to make his own scientific advancements without the previous work of other physicists, chemists, and innovators. Before we learn about Libby's life, we must give time and page to the lives of those who came first, building the path on which Libby walked and ultimately succeeded.

## MARTIN D. KAMEN (1913–2002)

Without the discovery of carbon-14, there would be no carbon dating method whatsoever. Martin Kamen played a huge role in that discovery, though his story, post-success, quickly became dark and troubled. Kamen was born in 1913 in Toronto, Canada, though he grew up in Chicago with his parents, who were Czech immigrants. In college, he studied chemistry, eventually earning his PhD in physical chemistry in 1936. Desperate to study chemistry at one of the country's most prestigious labs, Kamen traveled to the University of California, Berkeley, the home of the radiation lab and cyclotron. For the first six months, Kamen worked without pay. Eventually he was hired to oversee the cyclotron machine.

While stationed at the Berkeley laboratory, Kamen worked with Samuel Ruben to create the radioactive isotope carbon-14. At the time, the scientific community knew about the existence of carbon-14 but did not know that $^{14}C$ was naturally occurring on Earth. Instead, Kamen and Ruben used the 60-inch (152.4-centimeter) cyclotron accelerator to bombard graphite, producing carbon-14. While at the laboratory, Kamen was also able to confirm that the oxygen released in photosynthesis comes from water rather than carbon dioxide. This is the project that put Kamen and Ruben on the path to the $^{14}C$ discovery, as they were tracing the carbon-based molecules in the process of photosynthesis.

Directly following these major breakthroughs, Kamen was assigned to the Manhattan Project, a top-secret program, spread out among laboratories across the country, that sought to develop the science needed in order to create

The Oak Ridge National Laboratory in Tennessee was one of the main sites established as part of the Manhattan Project.

an atomic bomb. Kamen was located at Oak Ridge National Laboratory in Tennessee. Due to the secret nature of the project, employees had varying levels of information and security clearances. Some were left in the dark about what the project was hoping to achieve, while others had more details. This situation would soon spell trouble for Kamen.

According to Dr. Kamen's autobiography, *Radiant Science, Dark Politics* (1985), he was working on an experiment and asked a colleague to produce some

Dr. Martin D. Kamen fought allegations that he shared sensitive information about the United States' atomic program with spies from Russia.

radioactive sodium. When he opened up the container, it was glowing with a strange purple color, clearly extremely radioactive, much more so than would have been produced by the humble cyclotron. Suddenly, Kamen put the pieces together and realized that an atomic reactor must have been built at the laboratory in Tennessee. He blurted this out loud to Dr. Lawrence, a high-level researcher, along with Lawrence's army escort. Soon after, an investigation was begun in order to figure out who had leaked the news to Kamen.

Returning to his lab at the University of California, Berkeley, Kamen had a chance meeting with two Russian officials, undercover KGB (Komitet Gosudarstvennoy Bezopasnosti, translated in English as Committee for State Security) officers, at the party of a close friend. According to Kamen, one of the individuals asked him for help with an experimental radiation treatment for leukemia in order to help a coworker. To thank him, the three had dinner at Bernstein's Fish Grotto on July 1, 1944, which would become a fateful day for Kamen. Unfortunately, unbeknownst to Kamen, the FBI (Federal Bureau of Investigation) had been keeping a close watch. They saw him at the dinner with the Russians and immediately fired him from the California laboratory.

Over the next few decades, Kamen would continue to fight the accusations that he was colluding with Russians and knowingly supplying them with information about the American atomic program. In 1948, he testified to the House Un-American Activities Committee (HUAC). He was named as a suspected spy by the *Chicago Tribune* in 1951 and was denied not only academic jobs but also the issuance of a passport by the State Department. The pressure mounted, and Kamen eventually attempted suicide, though he was found by his wife, Beka, and saved.

Following the anti-Communist 1940s and 1950s, Kamen was once again able to obtain jobs in research and education. He was one of the founding faculty members of the University of California, San Diego, in 1961, and he continued to research photosynthesis into the later years of his life. In 1995, Kamen was one of the winners of the Enrico Fermi Award for lifetime achievements in energy research. Later, some would say that he should have received the Nobel Prize, as his discoveries were so important to the greater scientific community.

## SAMUEL RUBEN (1900–1988)

Kamen's scientific partner, Samuel Ruben, has a story that involves much less Communist intrigue, but it is nevertheless exciting and perhaps one of the best examples of the "American Dream" come to life.

Ruben was born in 1900 in Harrison, New Jersey. He went to public school in New York City, but, unlike his colleague, he did not pursue and complete a college education. He attended the Polytechnic Institute of Brooklyn, in New York, but withdrew after just a few years. However, Ruben was deeply interested in electronics, becoming a licensed ham radio operator early in life and then building his own radios from spare parts. While in New York, Ruben met a physics professor at Columbia University who took an interest in his talents, allowing him to sit in on lectures and use the physics library without being an enrolled student. With the help of that professor, Ruben set up his very own small, private lab in the mid-1920s, in New York City. He would conduct research there for the next sixty years.

Throughout his time at the lab, Ruben accumulated over two hundred patents for various inventions, many of which were technical in nature, hidden inside everyday electronics. He invented the dry electrolytic condenser, which is a part that was found in radios, television sets, and electric motors for decades. Each year, several hundred million condensers were produced. He also invented a dry-cell battery that was used in watches, satellites, and pacemakers. Batteries were his forte. During World War II, Ruben invented a sealed battery that maintained voltage despite any extreme changes in temperature. It would end up in walkie-talkies, mine detectors, and communications equipment used by the American military. During the war, Ruben turned over his patents to the government for military use.

Ruben would go on to publish multiple books about batteries and electrochemistry and receive many awards, though he was much more well-known for his contributions to electronics than he was for his work with carbon-14.

## SERGE KORFF (1906–1989)

It is worth noting, again, the accomplishments of Serge Korff, who laid the groundwork for the discovery that carbon-14 occurs naturally in the upper atmosphere. Korff was a physicist and pioneer of cosmic ray research who grew up in the Washington, DC, area. After graduating from Princeton University, in New Jersey, with his doctorate,

Dr. Serge Korff is pictured here with one of his cosmic ray antennas.

Korff worked for the Carnegie Institution, the Bartol Foundation, the California Institute of Technology, and the Mount Wilson Observatory. It was during this time that he began his cosmic ray research and charted the neutron bombardment of Earth's atmosphere. In the 1940s, he became a professor of physics at New York University.

In addition to widespread travel and experimentation, Korff found time to serve as the president of the American Geographical Society as well as president of the New York Academy of Sciences. He was a lifetime adventurer, venturing to the highest elevations possible in order to obtain data for his cosmic ray research. In 1937, while climbing a summit in Peru, Dr. Korff took photographs of a solar eclipse that were widely published. He passed away in 1989, at the age of eighty-three.

## WILLARD LIBBY (1908–1980)

When they think of the history of radiocarbon dating techniques, most people first think of Willard Libby. He is known as the father of the method and is credited with the discovery itself. Willard F. Libby was born in Colorado in 1908, the son of farmers. He attended elementary school in a two-room schoolhouse in Santa Rosa, California, and high school in Sebastopol, where he played tackle on his football team, all before moving to the University of California, Berkeley, in 1927. He received his undergraduate degree and PhD in chemistry at the University of California, Berkeley, while working on a California fruit ranch where he nailed together wooden fruit crates. This is how Libby paid his school tuition. He was tall at 6 feet 2 inches (188 cm), towering over others in the lab and on the farm.

Over the next few years, Libby worked as a professor of chemistry at the University of California, Berkeley. He worked primarily on the construction of Geiger counters, which measure radioactivity. Libby focused on counters that could measure weak or artificial radioactivity, specifically. An accomplished chemist, he was awarded a Guggenheim Memorial Foundation Fellowship in 1941 and hoped to work at Princeton University, but his work was interrupted early in the process.

It was World War II, and Libby felt called to work on the Manhattan Project, the secret program to develop atomic weapons. A day after the Japanese attack on Pearl Harbor in 1941, Libby volunteered himself to work under Harold C. Urey. Taking leave from Berkeley, the two met up at the Columbia University SAM (Special Alloyed or Substitute Alloy Materials) Laboratories, in New York City. While there, Libby worked on uranium enrichment, which is a process of isotope separation by which uranium becomes usable for nuclear power generation and military nuclear weapons. Today, uranium enrichment is heavily monitored and controlled worldwide in order to curb the development and construction of nuclear weapons. In the 1940s, during the Manhattan Project, the United States was working at a deliberate speed in order to enrich uranium as efficiently as possible. The United States government was hoping to construct and deploy the first atomic weapon in order to put an end to World War II. Eventually, with the dropping of bombs on Hiroshima and Nagasaki, in Japan, the United

This atomic bomb, named "Little Boy," was detonated over Hiroshima, Japan.

States would win the race to use this technology, for better or for worse.

During his time at the SAM Laboratories, Libby worked on teams that enriched uranium through gaseous diffusion and created new barriers to be used in uranium enrichment processes. On August 6, 1945, a bomb named "Little Boy" was dropped over Hiroshima, Japan. It contained enriched uranium. After the news of the bombing broke, Libby went home and told his wife that the bomb was what he'd been working on for the past few years. It had been a secret even to his own family.

During the 1940s and 1950s, Libby served on a myriad of boards and commissions, some associated with universities and others with national defense. He was a member of the Committee of Senior Reviewers of the Atomic Energy Commission (AEC) and the commission's General Advisory Committee, a committee he was appointed to by President Dwight D. Eisenhower himself. He also served as a consultant to the Office of Civil and Defense Mobilization, and he was a member of the Advisory Council to the Coordinator of Atomic Activities of the State of California. His work as a leader in chemistry and atomic defense cannot be understated. Libby was well-known and respected in his field. He was also a lifelong proponent of nuclear weapons. In a 1995 profile in *Time* magazine, Libby said of the bombs, "Let's build them as big as we can, and build all we can. Then war will become inconceivable."

Libby's second wife, Leona Woods, was one of the most notable women to work on the Manhattan Project. She was a nuclear physicist who studied under Enrico Fermi and Robert S. Mulliken. She was the only woman on a fifty-member team who built the world's first artificial nuclear reactor. She was present for the first nuclear chain reaction in 1942, and she oversaw the construction of neutron detectors. She also studied climatology, the structure of the universe, and nuclear power. She wrote two works of scientific nonfiction and a memoir titled *The Uranium People.*

It wasn't until after the war that Libby was able to publish his theories about radiocarbon dating. From there, Libby worked closely with the Eisenhower administration to defend their stance on atmospheric nuclear testing and to promote the president's Atoms for Peace program, which attempted to teach the public about the hopes of a nuclear future. While everyday Americans were horrified by the bombings of Hiroshima and Nagasaki and fearful about living in the nuclear age, Eisenhower was attempting to both calm fears and continue to press for peaceful use of nuclear technology. Atoms for Peace helped to create international discussion about the nonproliferation of nuclear weapons and gave political cover for United States nuclear weapons buildup and the arms race of the Cold War. Throughout the program, the United States exported enriched uranium to thirty countries, fueling research reactors. The first nuclear reactors in Iran, Israel, and Pakistan were built under the program as well.

Libby would continue to be an advocate for the peaceful use of nuclear technology throughout his entire life, though he resigned from the Atomic Energy Commission in 1959 in order to become a professor of chemistry at the University of California, Los Angeles (UCLA). He would retire from that position in the 1970s. Libby directed the UCLA Institute of Geophysics and Planetary Physics (IGPP) for over twelve years, during which the institute helped to oversee the Apollo space program and first lunar landings. Libby is also credited with starting the very first environmental engineering program at UCLA in 1972. Toward the end of his life, he became very interested in the environmental impact of emissions and air pollution standards. He remained professionally active until he died at UCLA Medical Center in 1980.

## OTHER CONTRIBUTORS

Serge Korff, Martin Kamen, Samuel Ruben, and Willard Libby are certainly the major players in the history of the carbon dating game, but their work would be useless without the contributions of several lesser-known scientists who offered their own unique gifts to the effort, some prior to the discovery, and some after.

## William Smith (1769–1839)

In order to even begin considering the idea of carbon dating, scientists and archaeologists first had to have a base understanding of the dating methods already in place prior to carbon-14 technology advances. William "Strata" Smith, who lived between 1769 and 1839, created the very first nationwide geological map. Today, he is known as the "Father of English Geology," but during his lifetime, Smith was ridiculed and ignored.

Smith was born in a small English village, the son of a blacksmith who died while William was just a child. Raised by an uncle and almost entirely self-educated, Smith found work as a surveyor and grew strong in his skills. His first experiences with rock layers were within coal mines owned by the Somersetshire Coal Canal Company. Smith was charged with inspecting the mines, and while he was there, he recorded the layers of rock and coal that were exposed by the mining process. These were some of the very first studies that formed the basis of stratigraphy, which is the geological study of rock layers and layering. Later, stratigraphy would become the bedrock of relative dating for archaeologists. Today, objects that cannot be dated are often still dated relatively using stratigraphy. It was Smith who discovered that the rock layers (strata) were arranged in predictable patterns and positions and that the stratum contained specific fossils. A talented artist, Smith drew tables and cross sections of the exposures he examined in quarries, mines, canals, road cuttings, and more.

In the early 1800s, Smith published the first geological map of Britain, which covered England, Wales, and parts of Scotland. The map was extremely detailed and colorful. It also included a geological cross section of the various strata. Unfortunately, the map did not sell well. Smith fell into debt and eventually ended up in debtor's prison. His home and property were seized, and he spent many years simply trying to survive as a surveyor. Though he was little celebrated throughout his life, Smith's work was foundational for geological surveys and dating techniques as well as the work of Charles Darwin. Today there are several buildings named after the man and a crater on Mars.

## Ernest Orlando Lawrence (1901–1958)

Ernest Lawrence was another innovative nuclear scientist who was born in the early 1900s. Though his life was cut short when he died at the age of fifty-seven, Lawrence's list of accomplishments is long. He founded the Lawrence Berkeley National Laboratory, home to so many scientists who worked on carbon-14, and he invented the cyclotron, which allowed Kamen and Ruben to synthesize the isotope.

Lawrence was born in South Dakota in 1901. He attended college for chemistry in South Dakota, and then he received his MA in physics from the University of Minnesota. He completed his PhD in physics at Yale University, in New Haven, Connecticut, where he remained as a researcher. In 1928, he moved to the University of California, Berkeley, to become a professor of physics.

Dr. Ernest Lawrence and his assistant prepare a cyclotron for its final testing.

During the early 1930s, he began to work with colleagues on the first versions of the cyclotron. The first one was only 4 inches (10.2 cm) in diameter.

He kept at the project, though, gradually making bigger and bigger machines that were able to help others with their research in high-energy physics. In 1939, Lawrence was awarded the Nobel Prize in Physics for his creation of the cyclotron. He was the first person from South Dakota to become a Nobel Laureate.

After World War II began, Lawrence, like many other physicists, became involved in research that would help the war effort. He worked at the Oak Ridge plant in Tennessee, and he was even an observer at the very first test of the nuclear bomb, the Trinity test, which occurred on July 16, 1945. Lawrence was a big supporter of the nuclear bomb effort and, after the bomb was dropped, felt proud that he had been a small part of the big project. Later, he was a proponent of the hydrogen bomb, a weapon even more powerful than the first nuclear bombs.

Sadly, while participating in a delegation to Geneva, Switzerland, to argue the case for the hydrogen bomb, Lawrence became ill and soon passed away. He was only fifty-seven years old. Less than a month after his death, two of the University of California's nuclear research sites were renamed after the man who had had such an impact on both the nuclear science community and the world at large.

## A Community Effort

Most scientific discoveries, though they may hinge on the work of one person, were not created in a vacuum. Discoveries are made within a context that includes decades—even centuries—of previous scholarship, work, and innovation. There are scores of unnamed researchers, fellows, students, and financiers who make the stuff of science happen.

One of those researchers is James Arnold, who was one of Willard Libby's close associates. He worked as a postdoctoral researcher under Libby to analyze samples with known ages in order to test the accuracy of radiocarbon dating.

While Libby won the Nobel Prize for this work, Arnold simply returned to Princeton University when the project was over. At Princeton University, he continued his own work on cosmic rays, using them to measure the age of lunar rocks, or moon rocks. He would also go on to be awarded the NASA Exceptional Scientific Achievement Medal in 1970 as well as found the University of California's California Space Institute in 1979. Arnold, for his part, was not involved in the Manhattan Project, preferring instead to focus his research on cosmic rays and lunar rocks.

It is the people like James Arnold who are the hands and feet of scientific discoveries. Though Willard Libby is the listed creator of the radiocarbon dating method, he would not have that title without the work of so many others who came before him, and his work would not endure without the scientists of today who continue to build upon his discoveries, ensuring that the technique remains relevant, efficient, and accessible to today's scientists. Even today the radiocarbon dating method continues to be modified as new discoveries are made.

# Marie Curie and Radioactivity

Marie Curie developed the theory of radioactivity and discovered polonium and radium.

Many of the major players in the discovery of radiocarbon dating were male scientists. But did you know that the person who gave a name to radioactivity was a woman?

Marie Skłodowska Curie studied in France, and she became interested in making precise radiation measurements. She would go on to conduct some of the most important and pioneering research on radioactivity. She won Nobel Prizes for her work, one in physics and another in chemistry. She wrote about her own work in a 1904 copy of *Century* magazine:

> It was at the close of the year 1897 that I began to study the compounds of uranium, the properties of which had greatly attracted my interest. Here was a substance emitting spontaneously and continuously radiations similar to Roentgen rays, whereas ordinarily Roentgen rays can be produced only in a vacuum-tube with the expenditure of energy. By what process can uranium furnish the same rays without expenditure of energy and without undergoing apparent modification? Is uranium the only body whose compounds emit similar rays? Such were the questions I asked myself, and it was while seeking to answer them

> that I entered into the researches which have led to the discovery of radium.

Even though she was a top student in her secondary school, Curie could not attend the University of Warsaw, in Poland, because it only admitted men. So, she continued her education in Warsaw's "floating university," which was a set of classes held in secret. Both Curie and her sister Bronya wanted to go abroad to earn an official degree, but they lacked the financial resources. Curie decided that she would work to support Bronya while she was in school, and Bronya would do the same after she completed her studies. For about five years, Curie worked as a tutor and used her spare time to study physics, chemistry, and math.

It was Curie who both developed the theory of radioactivity and coined the term. She also discovered both polonium and radium, naming the element polonium after her native country—Poland. She died July 4, 1934, from complications due to such heavy and constant radiation exposure throughout her life as a researcher.

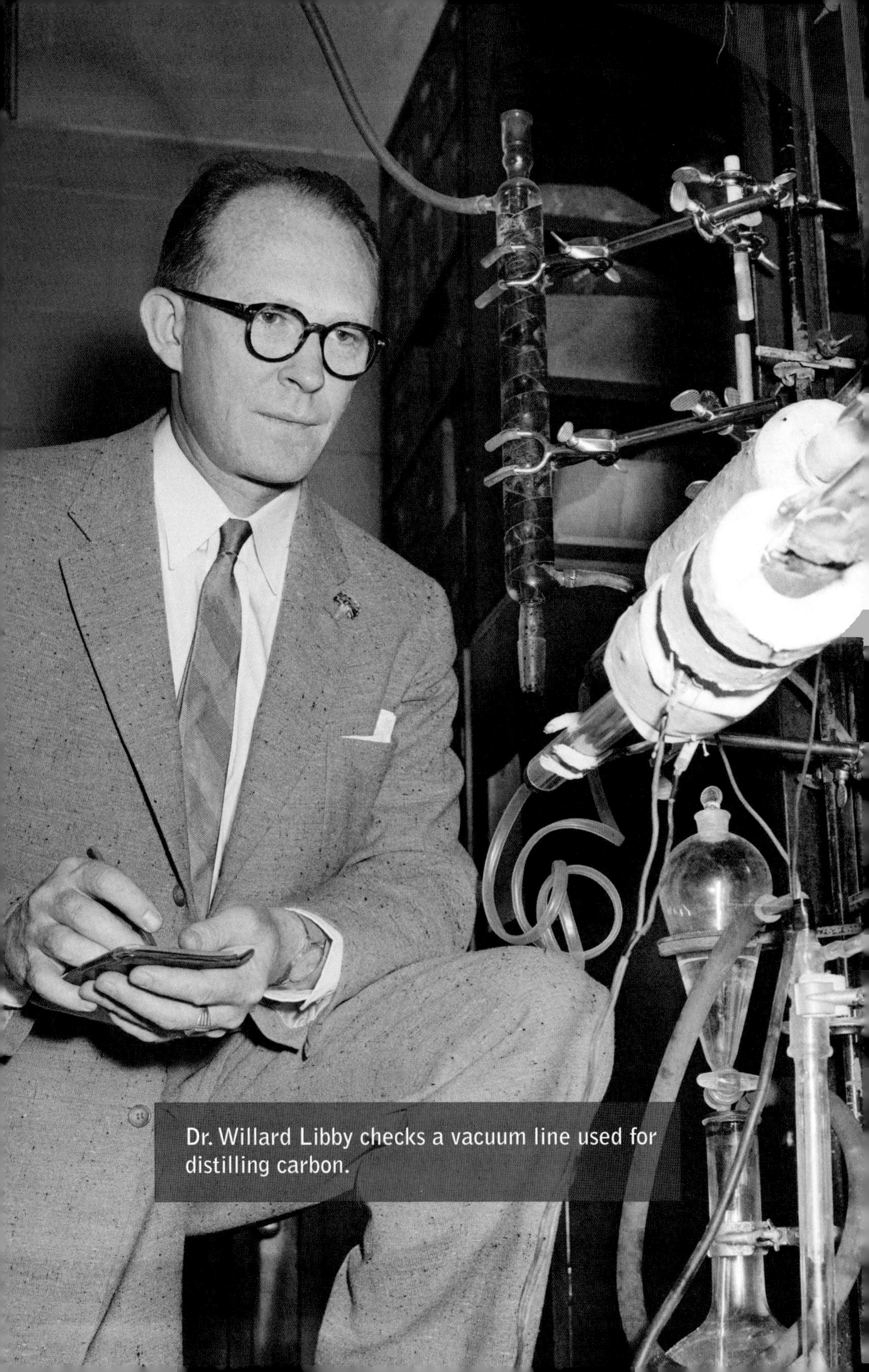

Dr. Willard Libby checks a vacuum line used for distilling carbon.

CHAPTER 4

# The Discovery of Carbon Dating

## TESTING A THEORY

While Serge A. Korff, Martin Kamen, Samuel Ruben, and several others paved the way for the discovery of radiocarbon dating, it was Willard Libby who is known as the father and creator of the method. During his time at the University of California, Berkeley, during World War II, Libby learned of Korff's research into cosmic rays and the production of secondary neutrons in the upper atmosphere. Thinking about this, he began to suspect that the isotope could be used for dating purposes. During these years, Libby was also heavily involved with the Manhattan Project to produce the first atomic bombs, but nevertheless, he kept up his interest in carbon-14.

Libby moved to the University of Chicago in 1945. This is where he would begin his work on radiocarbon dating. According to Libby, the research occurred in several stages. His early work culminated in the 1946 publication of a paper proposing that living organisms contain both nonradioactive carbon and carbon-14, the radioactive isotope. Libby worked with E. C. Anderson to determine whether living material was as radioactive as previously thought. At the time, there were no measurement techniques sensitive enough to detect extremely low levels of radioactivity; these techniques would come later. With Anderson, Libby utilized methane gas (transported from the city of Baltimore's sewage disposal plant, no less!) in experiments to discover whether samples could be enriched to show that they contain carbon-14. He contrasted this with other samples of petroleum-based carbon, drawn from deep below the earth, that was too old to show any radioactivity. The two researchers published a paper in a 1947 issue of *Science* in which they suggested that radiocarbon dating was a potential scientific possibility.

Next, Libby and his team collected samples from across the world, hoping to see that they had the same natural radiocarbon content even though they were found in such diverse locales as the North Pole and South Pole as well as places with high winds and ocean currents. They tested Norwegian spruce wood from Sweden, elm wood from Chicago, pine needles from Mount Wheeler in New Mexico, oak from Palestine, eucalyptus from Australia, and seal oil from the Antarctic. E. C. Anderson presented these results as his doctoral thesis at the University of Chicago. Their research was supported from a wide variety of sources: the University of Chicago, the Geological Society, the Guggenheim Foundation, and, interestingly, the United States Air Force.

After extensively studying the natural occurrence of carbon-14, they moved to the final stage of the initial research. Libby and Anderson joined with James R. Arnold to test their theory. With the help of the American Anthropological Association and the American Geological Society, mummified samples were collected for the research team's use. The oldest samples were pieces of wood found in the underground brick structures of two tombs from the First Dynasty in Egypt, those of Vizier Hemaka and King Zet. Relative dating and other methods put the tombs at about 4,900 years old, give or take two hundred years. They also tested wood from the upper chamber of a pyramid, a sample from the deck of a funeral ship found

Giant Redwood Tree Cut Down for Lumber in California, 1881

Giant redwood trees like this one can grow to be nearly 400 feet (121.9 meters) tall.

in the tomb of Sesostris III of Egypt, wood from an outer sarcophagus that had been buried underground, and even a piece of linen from the Dead Sea Scrolls. One of the most interesting samples they were provided was heartwood from an extremely large redwood tree known as the Centennia Stump. Felled in 1874, it had about 2,900 tree rings and was said to be 2,928 years old at the time it was cut, give or take fifty-one years. This sample would be particularly telling, as it put the radiocarbon dating method and tree ring–counting method head-to-head. Libby's method passed the tests, dating all of the samples relatively close to their previously given ages. He published the findings in a 1949 issue of *Science*. The report included a graph titled the "Curve of Knowns" that compared the known age of the sampled artifacts to the ages suggested by the radiocarbon dating method. Each of Libby's results lay within a very narrow range of the known ages. Over the next decade, more than twenty radiocarbon dating labs were founded across the globe.

## The RADIOCARBON REVOLUTION

Libby's discovery had an immediate and influential effect on the archaeology and geology communities. Some have referred to this fast and strong adoption of the method as the "radiocarbon revolution." For so many years, investigators had relied on relative dating that used comparisons to draw conclusions. Libby's technology provided the very first method that was meant to be objective. Numerical dates could be attached to organic matter. What's more, the dating technique actually began to revolutionize previously held theories that were believed not only by scholars but by everyday people. For example, as archaeologists began to test the ages of artifacts from both the Eastern and Western Hemispheres, it became clear that civilization did originate in Europe, as was the prevailing theory. This "fact" was pervasive, solidifying, for many people, the idea that those born outside of the European center were somehow less advanced or intelligent. Radiocarbon dating began a breakdown of this theory that slowly led to a much less Eurocentric understanding of our global community. In addition, Libby's own analysis of wood samples taken from trees that were once buried beneath glacial ice gave a new date to the end of North America's last ice age. Scientists had long believed that it ended twenty-five thousand years ago. Libby's analysis suggested that the last ice age had ended only eleven thousand years ago.

When radiocarbon dating was first introduced, Libby believed that there was a threshold of twenty thousand years for the measurement of ages using his method. But scientists have kept working to improve the technology, and today the method can very reliably date materials that are as old as fifty thousand years. Of course, it can date objects that are even older, but fifty thousand is the point where reliability and accuracy begin to waver. Time will tell if there will be new advancements. For now, it seems as if radiocarbon dating is here to stay and will be continually improved upon. In 2016, the radiocarbon dating method was recognized as a National Historic Chemical Landmark by the American Chemical Society.

## NEGATIVE REACTIONS

Of course, not everyone is a huge fan of carbon-14 dating. Ever since its inception, there have been arguments within Christian circles about the reliability and accuracy of radiocarbon dating, especially as it relates to the age of Earth. So, how old is Earth? This may seem like a simple question, but it gets complicated when you consider modern scientific technology together with age-old religious texts and traditions.

For those who take the Bible literally, the account of creation in the Book of Genesis, as well as genealogical records provided throughout the Bible, indicate that the world is only about six thousand years old. This number pales in comparison to scientific determinations that date the world to be a little over four *billion* years old. The four billion number has been reached due to years of radiometric dating, which is somewhat of an umbrella term, under which radiocarbon dating resides, along with potassium-argon dating and uranium-lead dating. These dating methods are very closely related to radiocarbon dating, but they utilize substances other than carbon-14 to date absolute age. It is important to remember that radiocarbon dating is just one of the radiometric methods used to date the age of Earth and its artifacts.

Over the years, radiometric dating methods have caused some who believe in a literal account of the Bible to reevaluate their readings and question the relationship between science and religion. While some people and groups have found that radiometric and radiocarbon dating and a literal account of the Bible can coexist, others disagree. One can find websites, books, and more dedicated

Some groups claim to have found and carbon-dated wood from Noah's Ark, a ship described in the Old Testament.

to disproving the science behind carbon-14 dating in order to prove the biblical account. These analyses can be long and complex. Some even suggest that the great flood as accounted in the Book of Genesis could have buried massive amounts of carbon in Earth, disturbing the known carbon ratios and changing age estimates to more accurately reflect the biblical narrative. Many others have taken a hybrid approach, fitting what we know of science into what we know of scripture.

## AN IMPORTANT TEST

We know that radiocarbon dating has been fully embraced by some and passionately rejected by others. Nowhere can this division be more starkly seen than in the debates over the Shroud of Turin. The shroud is a linen cloth that has traditionally been associated with Jesus, specifically with his burial. As something that is said to have touched Jesus, it is known as a relic and has been incredibly important to many people. The length of cloth bears the image of a man who is said to be Jesus. The item's name comes from where it is stored, in the royal chapel of the Cathedral of Saint John the Baptist in Turin, which is in northern Italy. It measures about 14 feet by 3 feet (4.3 meters by 0.9 meter). The shroud dates back to the mid- to late 1300s, but its history is complicated and often contested within various church circles. It was transferred to Turin in 1578, where it has remained ever since.

Since the shroud is an object of both historical and spiritual importance, people were quick to suggest that it be radiocarbon dated once the technology became available,

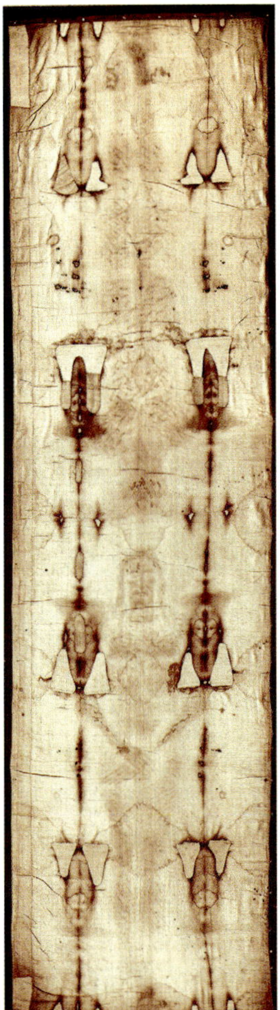

This full-length image shows the entire Shroud of Turin, which some believe bears the image of Jesus of Nazareth.

but others were just as quick to oppose it. While radiocarbon dating was first suggested in 1960, the technology was so immature at the time that a very large sample of the fabric would have been needed in order to do a proper test, which would have damaged the relic and the images on it. In the 1970s, as the technology improved, the Catholic Church, who owns the shroud, officially founded the Shroud of Turin Research Project (STURP), which may be one of the only formal groups brought together in order to decide if an object should or should not be radiocarbon dated. The group approached several candidate laboratories in order to choose not only a laboratory, but a series of tests that would be performed on the object. Unfortunately, by the early 1980s, the STURP group and the laboratories were in disagreement. The laboratories wanted to do radiocarbon dating as the primary test, to be completed no matter what, while the group was interested in the performance of other radiometric examinations to occur prior to the radiocarbon test.

At last, an official meeting was held in Turin, Italy, between representatives from the labs and those from the Church. They drew up an official list of protocols, which included but were not limited to the following: Carbon-14 dating was to be the only test performed, and seven laboratories would perform independent tests. Each laboratory would receive about 9 square centimeters (1.4 square inches) of cloth as a sample, and labs would not communicate with each other during the tests. Then, as the year wore on, the Church countered with new protocols, deciding that fewer laboratories would take part. This conversation, carried out both in the press and in private, was a long ordeal of back-and-forth between the Church hierarchy and leaders of the scientific project, mostly surrounding the locations of the tests and the number of times the shroud would be tested. Some came to think that the test would never really happen.

After ten years of lively debate, the final protocol was decided upon and released. In April 1988, it was decided that three laboratories would perform the test: Oxford, Great Britain; Tucson, Arizona; and Zurich, Switzerland. The laboratories would not communicate with one another, and their tests would be filmed. All the samples would be taken from the same portion of the cloth.

On April 21, 1988, samples were taken from the shroud in the presence of priests, the cardinal, a fabric expert, a representative of the test equipment, a camera operator, and a photographer. The tests were performed in May, June, and August of 1988, and their results were communicated to the British Museum, which acted as a neutral party during the entire process. On October 13, 1988, the official results were announced by Cardinal Anastasio Ballestrero, dating the shroud to the years between 1260 and 1390 CE. This would place the Shroud of Turin within the medieval era, which closely matches the first appearance of the shroud in church history. In the end, the radiocarbon dating process said nothing about the image on the shroud (as that was not what was being tested), but it does seem to have disproved the account of the shroud as a burial cloth for Jesus. The radiocarbon results have since been disputed on the grounds that they might have been affected by contamination from cloth fibers used to repair the relic when a fire damaged it in the Middle Ages. In 2013, Giulio Fanti, a professor of mechanical and thermal measurement at the University of Padua, in Italy, published a book, *The Mystery of the Shroud*, that called into question the 1988 carbon-14 tests on the basis that they were compromised by the fact that the testers examined an area of the cloth that was patched with an interweaving that the tests couldn't detect. New technology suggests the cloth could be dated between 200 BCE and 200 CE.

## MAKING CORRECTIONS

Debates between scientists and religious groups have ensued over the years with very few conclusions reached. Both continue to grow and evolve and ideally find ways to do so in comfortable community with one another. Even for those scientists who are or have been head over heels for radiocarbon technology, it is true that the carbon-14 dating

These researchers are taking a look at tree rings, which remains a popular way to date trees and learn about Earth's past.

technique has never been perfect. From advancements in the reliability and accuracy to related forms of radiometric dating that have been developed over the years, science is constantly changing and readjusting. It is never perfect or static.

To further explore this point, we return to the subject of calibration. In its most simple definition, calibration takes the "radiocarbon years" produced by the dating method and turns them into calendar years that are easier for us to understand and utilize. This process is necessary because the ratio of carbon-14 to carbon-12 has changed over the years. As early as 1955, Libby was suggesting that this ratio could be variable, but it was years before formal corrections were applied and the calibration curve was created.

The first curve was created after many tests of tree rings that provided information that dipped deeply into the past. In the 1960s, Wesley Ferguson created the very first sequence of tree rings to be used for this purpose, using bristlecone pine trees. Believe it or not, he was an academic at the Laboratory of Tree-Ring Research (LTRR) in Tucson, Arizona. The lab was founded in 1937 and thrives today in a nearly new building that contains laboratories, office space, and a wood sample archive. There are many tree-ring labs around the world, but the LTRR was the first of its kind.

The data collected by Wesley Ferguson at the LTRR was then used by Hans Suess, an Austrian chemist and physicist, to make and publish the first calibration curve in 1967. The curve contained a straight line that showed a basic comparison of radiocarbon age to calendar age. It then displayed another line, a little squiggly and wavy in shape, that included fluctuations in the $^{14}C$ to $^{12}C$ ratio. This first curve assumed that the variation was global in nature.

Since 1967, many calibration curves have been created and published, using many different data sets. Studies of tree rings, coral, layers of sedimentary rock, and other materials have been undertaken in order to get more concrete data on the variations. Today, there are separate calibration curves for the Southern and Northern Hemispheres as well as for the oceans. Curves are useful for translating raw data into usable data. When a test produces a sample age in radiocarbon years, a researcher can go to the calibration curve and determine an age range based on calendar ages. Because the amount of carbon-14 in the atmosphere has not remained constant over time, you might see a difference like this: A plant with a radiocarbon age of 5,000 years old could be about 5,700 *calendar* years old, and 20,000 radiocarbon years is about equal to 25,000 calendar years. Do you see why the corrections are important? Without the calibration curve, we would have a vast misunderstanding of the true age of ancient organisms.

## RADIOCARBON DATING BECOMES the STANDARD METHOD

Soon after Libby's initial discovery, radiocarbon dating took on a life of its own, suddenly utilized by scientists across the globe for the efficient, accurate dating of all kinds of ancient organic material. Labs were built, and the technology continued to be improved upon, year after year. As the technology became less novel and more regularly acknowledged as standard practice, the lens through which people viewed the dating method was able to grow.

Ancient Egyptian sites, such as this step pyramid at Saqqara, have long been of interest to archaeologists and other researchers.

Instead of simply looking at a single object and using carbon dating to reveal how many years ago that object lived or was created, scientists began to use the method for more expansive and far-reaching projects, such as the mapping of people groups across the planet or the use of dating to determine how poachers were interacting with the current ivory market. Carbon-14 isn't just useful for putting a date on an object. It can help us form entire cultural worldviews and act as a tool to help solve modern-day problems.

In recent years, the method has also been utilized not only to chart the ages of items that have never been accurately dated, such as the Shroud of Turin, but to verify some of our standard beliefs about ancient history. A study released in 2010 showed that, although radiocarbon has been used to date ancient Egyptian artifacts since its very inception, today the technology has improved so much that scientists are able to confidently do checks on long-held chronologies of ancient history.

For centuries, experts have used relative data and so-called "floating" chronologies to ground their chronological understandings of ancient Egypt. The dates have always been the subject of heated debate, as the only available data included Egyptian records, such as inscriptions at sites (for example, Saqqara and Karnak), and writings from historians of the third century BCE. These dates are considered "floating" because they make sense side by side with each other but are not anchored onto any absolute dates.

Here's what dating looked like for ancient Egyptian sites before radiocarbon dating. The oldest known pyramid, the step pyramid at Saqqara, also known as the Pyramid of Djoser, is made of limestone and stands about 203 feet

(61.9 m) tall, a towering shape in the Egyptian desert. We know from writings that the pyramid was built as a tomb for King Djoser. Historians usually date his reign from 2667 to 2592 BCE.

Djoser was the first king to rule within the Third Dynasty. He was also the first king to build using stone. For many years, historians felt sure of the date of Djoser's reign, but a recent paper used astronomical calculations to date the reign, pushing it about seventy-five years later than the usual dates. This contradiction remained for some time, as radiocarbon dating was too imprecise. With a possible error range of one hundred to two hundred years, carbon dating could not be used to solve the problem.

However, a team at the University of Oxford, in the United Kingdom, recently developed a new strategy and made a new attempt to date the tomb using radiocarbon dating. Using their new strategy, researchers combed museums around the world to find plant remains (which would contain carbon matter) that were directly associated with reigns of particular kings.

They collected 211 plant samples, many of which were from other pyramids where kings were buried. These samples, which included seeds, plants, and papyrus, were then carbon dated. By this time, carbon dating technology had become so advanced that the team only needed samples as large as a grain of wheat in order to conduct the studies. The samples were dated using Oxford's accelerator mass spectrometer.

Then, the researchers, who came from the United Kingdom, France, Austria, and Israel, used a mathematical model to compare patterns in the radiocarbon and historical dates. They made three separate models for the three periods—Old Kingdom, Middle Kingdom, and New Kingdom.

Separating the periods like this shrunk the margins of error to seventy-six, fifty-three, and twenty-four years, respectively. The team found that the Old Kingdom, including Djoser's reign, began between 2691 and 2625 BCE, which is earlier than the previously proposed dates. The dates the team found for the Middle and New Kingdoms were also earlier than expected.

Though the dates found using the radiocarbon dating method were slightly earlier than expected, they were not groundbreaking results. In fact, their main contribution was confirming that scientists had previously been on the right track with their data from the relative dating methods, such as stratigraphy efforts and the analysis of ancient inscriptions and texts. If much of the assumed chronology was correct, what does that mean for dating methods going forward?

Essentially it suggests that while multiple types of dating methods can and should be utilized when attempting to date ancient artifacts and locations, radiocarbon dating has reached the point of standard-bearer. It is an efficient and accurate option that is almost clinical in nature—it doesn't require scholars who are trained to distinguish handwriting or interpret the strata of sedimentary rock. All one needs is organic material that contains carbon-14 and an AMS machine on which to test the sample. Data obtained from radiocarbon dating can then be back-checked with other methods, rather than doing the experiments the other way around, with carbon dating as the checker.

Technology has come such a long way since the early 1950s, and we are now able to put accurate dates to artifacts that hit within a short range of decades. Of course, the method isn't perfect. There are still margins of error, and some issues still stand. When it comes to the tests on Egyptian time periods, archaeologist Hendrik Bruins pointed out something that could be a major issue in the accuracy of the dating—the eruption of the volcanic island of Thera in the Aegean Sea.

Carbon dating tests have attempted to date the eruption to one hundred years prior to the beginning of the New Kingdom, which would be about 1570 BCE. However, other radiocarbon and historical dating has suggested that the volcanic eruption definitely took place *during* the New Kingdom. The jury is still out on this one. All we know for sure is that radiocarbon dating has improved enough over the last seventy years to become the standard within scientific and archaeological circles.

## Do You Want a Radiocarbon Analysis?

Today, scientists who are hoping to have samples dated using the radiocarbon method are not limited to the University of California, Berkeley. There are laboratories all over the United States, Canada, and the rest of the world that are happy to provide the service—for a fee.

At the National Ocean Sciences Accelerator Mass Spectrometry (NOSAMS) facility in Woods Hole, Massachusetts, carbon-14 analysis is done primarily for the ocean sciences research community. Samples sent here might include parts of shells, wood, or coral. The National Ocean Sciences Accelerator Mass Spectrometry facility accepts samples from research laboratories, so you probably would not be able to just walk up to the facility with a seashell you found on your beach vacation.

Scientists from various qualified labs, however, are met with a complete menu of options when they approach NOSAMS about dating their particular samples. In fact, there are facilities to process nearly any carbon-containing material, including water.

The lab has two different accelerator mass spectrometry machines, among other types of equipment. A sample sent to NOSAMS can take anywhere from three to six weeks to return to sender, on average. A request for faster turnaround can be made by simply submitting a statement of scientific justification explaining the urgency of the project.

Samples at the NOSAMS lab have a range of prices based on the type and number of analyses to be completed. A bone sample, at full price, costs around $1,200. But discounts are given for the federal government and academic research, bringing the cost of a bone sample down to $600. At the lower end, a sample of graphite can cost about $200, while carbon dissolved in water can run anywhere from $700 to $1,000.

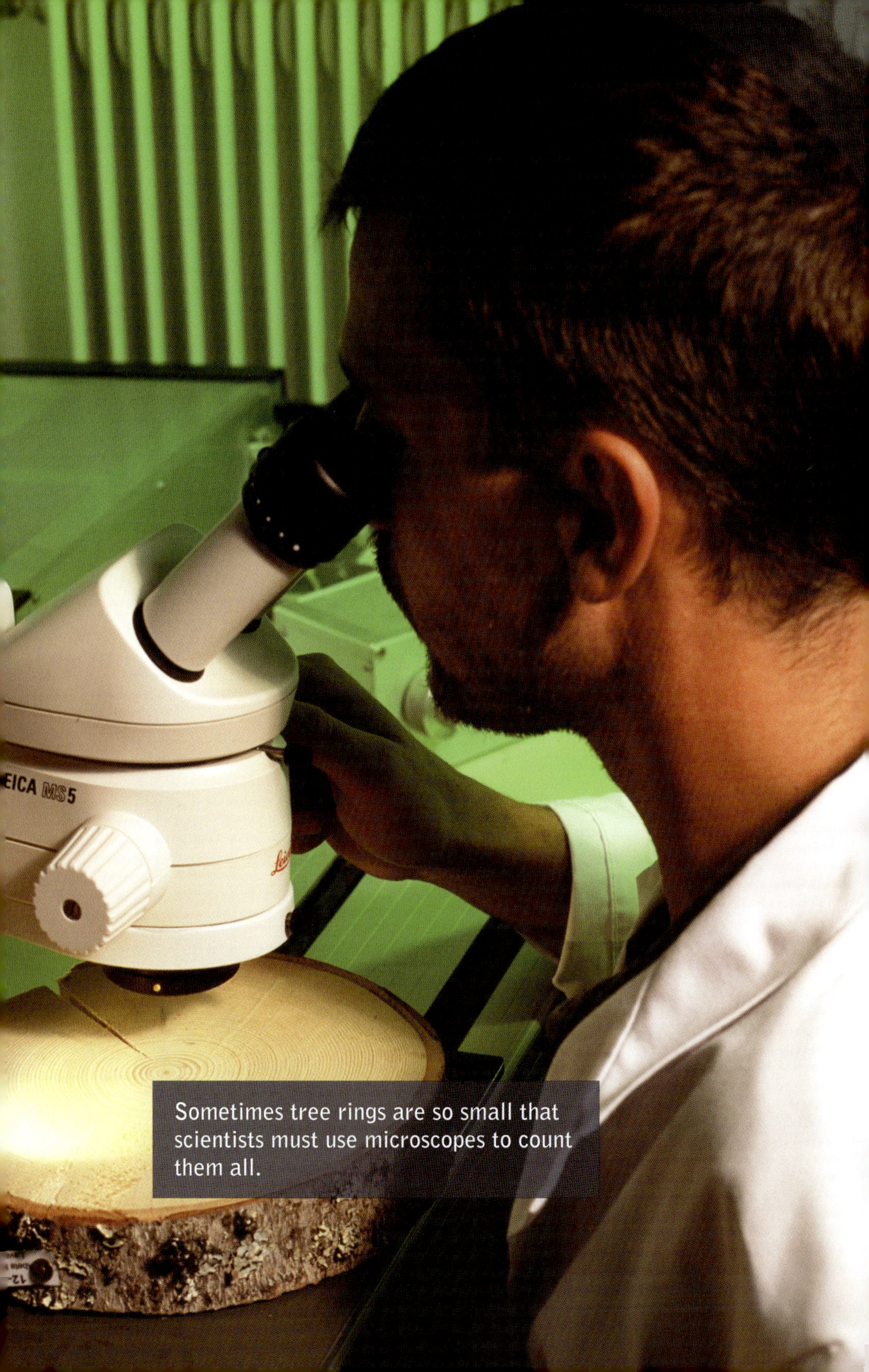


Sometimes tree rings are so small that scientists must use microscopes to count them all.

CHAPTER 5

# Carbon Dating Today

Though the technology behind radiocarbon dating was created decades ago, the method is still in widespread use today, and it continues to be the standard practice for archaeologists and researchers around the world who are hoping to attach dates to their discoveries, increasing the relevancy of their finds and adding context to history. Radiocarbon dating has been used in recent years to attach date ranges to mummies, sharks, religious texts, oysters, canoes, soil samples, and even bodies of water. We are going to take a look at a few of these examples as well as some competing dating methods and a few carbon dating problems that have come up in the last few years. Now, more than ever, it is important that we understand all the ways in which we can utilize this technology to make sense of our world, especially as the climate changes and history continues to move further behind us.

## DEAD SEA SCROLLS

Remember those Dead Sea Scrolls that paleographers analyzed? After Willard Libby performed his own test on a piece of related linen, decades passed. Libby wasn't able to test the scrolls themselves at first because the technology of the 1950s required large samples. It wasn't until 1991 that researchers were able to carry out radiocarbon dating on the Dead Sea Scrolls. Robert Eisenman and Philip R. Davies carried out a series of tests in an AMS laboratory in Zurich, Switzerland, on samples from fourteen of the scrolls. These tests were seen as a necessary step, as critics had long challenged the actual antiquity of the scrolls, believing them to be forged. The scrolls themselves, which comprise a group of more than eight hundred, only refer to historic events in two of the scrolls and thus could not provide conclusive evidence of a finalized date range.

The 1991 tests measured the amount of carbon-14 remaining in the scrolls, which told scientists that they were first written between 2,000 and 2,200 years ago, dating the scrolls between the second century BCE and the beginning of the first century CE. This is close to the paleographic estimates, which judged the scrolls to be written between the third century BCE and about 68 CE. At the time, the margin of error was known to be only about thirty to sixty years. All of the suggested dates line up with the theory that the scrolls belonged to a Jewish population that lived in the area until Roman troops destroyed the settlement around 70 CE. Another test was conducted between 1994 and 1995 in Tucson, Arizona, using the University of Arizona's

AMS facility. The tests, which were done on twenty-two scrolls as well as a piece of linen, ended with similar results.

The Dead Sea Scrolls, one of the most significant archaeological finds in human history, were well-served by radiocarbon dating, especially by accelerator mass spectrometry. Without these tests, the dating would be left up to handwriting analysis and Libby's first attempts at radiocarbon dating, which produced inaccurate results that differ from all of the other, more recent tests. For sensitive religious documents, carbon dating provides a window into the historic past of the objects, lessening the noise around conversations of their authenticity and age and allowing people to refocus on their content as well as the context in which they may have been written and saved.

## MYSTERY SOLVED

A more recent example of the ways in which radiocarbon dating can help to uncover mysteries of the past is the story of a forty-eight-year-old discovery that remained a mystery, until now. A man named Larry Reeve liked to fish on Pigeon Lake in Ontario, Canada. On a warm day in May 1970, Reeve, who was only twenty-nine years old at the time, noticed something in the water. He thought it was a log on the lakeshore, but he quickly noted its unusual shape—the log was dug out to be a canoe. Since it was the 1970s, Reeve did not call in a licensed archaeologist. Instead, he hauled the canoe back to his family's cottage, using his boat to drag it. The canoe was so filled with water that it was impossible to bring it completely out of the lake. It took about two full

## Climate Change and Carbon Dating

In the last couple decades, the conversation around climate change and global warming has been heating up, so to say. While this topic might be more at home in an environmental studies book, archaeologists and historical researchers are now becoming concerned about the effect that climate change could have on the continued accuracy of carbon dating methods.

According to writer Adrienne LaFrance, fossil fuel emissions are increasing at dangerous rates, and those emissions could skew the radiocarbon age of organic material that is created today. When fossil fuels are burned, extra carbon dioxide is released into the air, diluting the amount of radiocarbon in the atmosphere and thus the amount being absorbed by organic material. In an article for the *Atlantic*, she explained:

> If Graven's calculations are correct, carbon dating as we know it today will no longer be reliable by the year 2030. Which means scientists won't be able to use carbon dating to distinguish between new materials and artifacts

that are hundreds or thousands of years old.

Even though today's carbon-14 dates can vary between thirty and one hundred years from the "real" age of the particular material, changes in the atmosphere that are happening due to carbon emissions could require some extreme amounts of adjustment.

All of the talk of radiocarbon dating complications has archaeologists, historians, physicists, and other scientists worried. They are used to some ambiguities and the need for adjustments. But it will be troubling if researchers cannot distinguish an organic material from 2050 from something from the eleventh century, all due to climate change.

Carbon dating has been used to date items like wooden boats, canoes, and tools.

years of tugging, little by little, until Reeve was finally able to pull the canoe out.

After rescuing the boat from the lake, the young man decided to donate his find to the Peterborough Museum and Archives, in Ontario, Canada. It has remained in the museum ever since 1975, when it arrived. Reeve knew the boat must have historical significance, but he didn't know exactly when it was crafted or for what it may have been used. The museum staff, who were knowledgeable about the local area, were able to note that it came from the Michi Saagiig people of the Anishinaabe Nation, though its age remained a mystery.

Finally, in November 2017, Reeve and his friends at the museum had the answer they had always wondered about. A local archaeology professor had taken an interest in the canoe and, noticing some loose wood on its body, asked if he could run a radiocarbon analysis at the Lalonde Laboratory at the University of Ottawa, in Canada. After running the test, it was determined that the canoe was built between 1732 and 1807, before European settlers entered the region. This matches the oral histories of First Nations people in the area. First Nation elders had long spoken of collecting rice and fishing on Pigeon Lake. The canoe adds physical evidence to those stories, which have been passed down orally for hundreds of years. Without carbon dating, the canoe may have remained a mystery, destined to live in dusty archival space within the Peterborough Museum. Today, paired with information about the oral histories and carbon dating evidence, it is on display for visitors to see and discuss, dated evidence of those First Nations peoples who lived in the region prior to European settlement.

## OLD SHELLS FULL of NEW INFORMATION

It is important to remember that radiocarbon dating methods aren't only useful for gleaning information about man-made objects, such as boats and texts. These same techniques can also be used on natural organisms, like oyster shells. For the past few years, researchers have been working to dig up shell middens around both shores of the Chesapeake Bay. These middens are primitive landfills that hold ancient debris, especially great amounts of oyster shells, suggesting that ancient peoples living in the area relied on oysters for a large part of their diets.

Recently, scientists were able to sift out some shells from within the middens and send them out for carbon dating. The earliest shell found at the site was dated to the Early Woodland period, going back as far as 1200 BCE. This indicates that between the Early Woodland period and European settlement, oysters were a primary food source. This is valuable information, and it was retrieved not a moment too soon; due to changes in water levels, shell midden sites like this one will most likely be underwater in the next ten to twenty years. It is imperative that researchers send samples for carbon dating now, before the organic matter is lost to the open water.

## AN EVOLVING TECHNIQUE

If scientists had left radiocarbon dating techniques just as they were when Willard Libby discovered the possibilities within the isotope, carbon dating would be much

Shell middens like this one were used by Native Americans as trash piles for animal bones, shells, and other refuse.

less accurate. In order for the science to stay relevant, it needs to change with the times, as new information becomes available. And these days, it's doing just that. The method has become more precise and reliable over the course of the last several decades. One of the newest innovations is an update in the sample size—rather than dating bulk material, even of a small physical size that could contain several molecules, some scientists are taking an even more miniscule approach.

At the Oxford Radiocarbon Accelerator Unit (ORAU), a laboratory in the United Kingdom, the team has been extracting the amino acid hydroxyproline from bone collagen. They chose this amino acid because it carries a heavy load of the carbon in mammal bones. The extraction and dating of such a specific and small sample is known as single compound AMS dating. The method has been in the works since the 1980s, but it has never been wholly successful, at least not until recently. The team at ORAU says that its findings have been proving both accurate and reliable. Their work has also raised new questions in the archaeological and scientific communities, as they recently dated Neanderthal remains from Croatia to a date that is significantly older than previous tests. While ORAU's compound-specific dating method is currently only available for projects within the Oxford laboratory, the team hopes that it will someday be commercially available to other researchers around the globe.

This archaeologist found animal bones, nails, and ceramics on a dig in a public park in Baltimore, Maryland.

## PUTTING the DATES TOGETHER

Ever since radiocarbon dating became a popular technique in the 1950s, scientists have been using the method to assign dates to all sorts of objects that are found in our world. Ancient bones, texts, wooden artifacts, shells, and other materials are quickly and reliably dated using AMS, and then historians and archaeologists are better able to tell the full story of an item and its location. These days, researchers in Canada are pooling the available records together in order to map migration patterns and answer some major archaeological questions.

In 2014, the Canadian Archaeological Radiocarbon Database (CARD) began an expansion that had been planned for years. CARD was started in the 1980s, and today it holds over seventy thousand radiocarbon records hailing from seventy different countries. While there are other radiocarbon databases in the world, CARD is definitely the largest, and its ambition is global—it is working to expand its software into an open source format, making it easier for researchers to both contribute to and access the wealth of information that the database contains.

The workers at CARD are preparing for an uncertain future. Due to environmental changes and constantly evolving technology, many items that have already been dated may need to be re-dated in the coming years and decades. At CARD, the centralized database hopes to include comprehensive information about each test that has been completed, making it simple for students and researchers to study, learn, and build upon previous work. Some experts are also hopeful that such a large, organized data set could help researchers make better population estimates and trace human populations as they moved across the earth over time. With properly dated finds that are categorized efficiently, scientists can ask bigger questions about large-scale changes that have occurred over decades, centuries, and even thousands of years.

In 2015, specialists at CARD used the data system to make the very first continent-wide map of human occupation within the Americas, spanning the past thirteen thousand years. Some at CARD are hoping to use this map to match up traditional indigenous stories about population movement, wars, and relocation with patterns that are clear from the carbon dating map. Is this an easy task? Not a chance. Some radiocarbon dating information must be kept secret as to avoid looting. If everyone is told where an ancient relic or valuable object resides, it could easily be stolen. Additionally, it may prove difficult to obtain permissions from all of those who have completed carbon dating projects across the planet over the last seventy years. Finally, but unsurprisingly, it is difficult to find funding for an archive of this scale, though keeping the archive open-source will help the project continue to obtain data at minimal cost.

## The NEXT FRONTIER: DNA DATING

With the discovery and successful test of radiocarbon dating in 1950, archaeologists suddenly had a new tool that would make their jobs easier and open up windows to the past that we may never have imagined. Ancient objects were shown to be older or younger than previously thought, while others had tests that matched perfectly with previous theories.

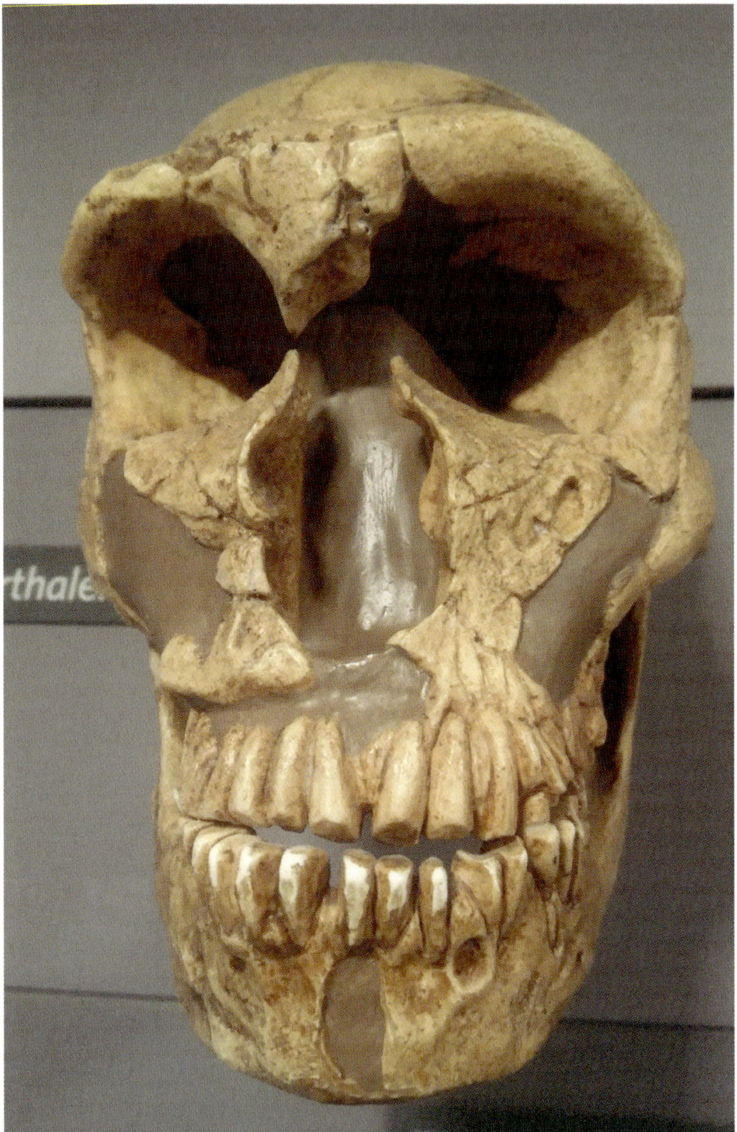

DNA has been recovered from Neanderthal fossils like this skull.

Since the discovery, carbon-14 dating has allowed scientists to gain insight not only into historical stories about our collective past, but into ideas about how we can and should be living our lives as we move forward. How can we preserve food supplies, like ancient peoples living in the Chesapeake Bay area did with their oysters? How can we better understand the role of power and empires in early pre-Christian communities? How can an object add weight and meaning to the oral histories of an oppressed people group?

These questions, and so many more, are kicked up by the explosion of information that was suddenly at our fingertips when carbon-14 dating methods emerged. Today, scientists continue to search for more and better ways of understanding and dating our world. One of the newest and most promising ideas is DNA analysis.

Ancient DNA, for the most part, has not been preserved. It is not kept perfectly sealed in a tiny vial, ready to be checked for clues to its makeup. Instead, ancient DNA is often collected from mummified tissue, skeletal material, plant remains, ice cores, medical specimens, and other things.

Scientists first started to understand the promise of this DNA in 1984, when colleagues at the University of California, Berkeley, reported that traces of DNA from an ancient zebra-like species, the quagga, had remained in the specimen over 150 years after the death of the animal and that the DNA could be extracted and sequenced. Soon after, researchers further discovered that mummified human samples contained DNA that could be extracted and sequenced. In the 1980s, DNA sequencing was not in its prime, so experiments took a painfully long time. However, the field eventually began to progress rapidly alongside strides made in the general field of DNA research.

As you may know, all living things contain DNA (deoxyribonucleic acid), a chain of nucleotides. DNA contains biological information that is used in the growth, development, and functioning of organisms. We all have DNA, though it is extremely small and difficult to understand.

Every tiny strand of DNA, which is invisible to the naked eye, contains more information than we can even begin to imagine. DNA changes over time, collecting historical information as it goes. So, when it is used carefully and studiously, DNA can be used to understand the evolution of modern humans, to identify individuals, and even to trace the migrations of people, plants, and animals across the earth. Some anthropologists are even using DNA as a useful dating method in places where carbon-14 just won't work as well.

Take the example of a cave in present-day Belgium. About fifty thousand years ago, a Neandertal relieved himself there, depositing a sample of his DNA. In the soil, the urine mixed with minerals, allowing traces of the DNA to remain embedded in the cave floor. Earth that fell from the cave's ceiling helped to further cover the sample and keep it protected for tens of thousands of years. Now, scientists are digging up the DNA and putting it to use.

In 2017, a report was published that detailed the Neandertal DNA that geneticist Matthias Meyer and doctoral student Viviane Slon discovered in cave sites from Spain to Russia. In one Siberian cave site, the pair found DNA from layers of sediment where no bones had ever been found. Before this discovery, historians understood the period of human occupation in the cave to be tens of thousands of years earlier than what the DNA evidence showed.

Radiocarbon dating can help date bone samples, but it can do nothing without that carbon-containing sample. However, radiocarbon-dated bones did help to corroborate the story that was told with the soil DNA, making the researchers more confident that the new technology was reliable. Now that DNA can be extracted from soil and sequenced, providing information about the organism that used to be there, scientists will have to depend much less on bones, hair, and similar physical samples. While DNA dating likely won't altogether replace the carbon-14 method, it certainly adds to the work that radiocarbon testing has set out to do, helping to give even clearer and larger pictures of historical events.

# Chronology

**1769** William Smith, who created the first geological map of England, is born.

**1869** Russian chemist Dimitri Mendeleev begins to develop the periodic table of elements.

**1886** French physicist Antoine Henri Becquerel discovers radioactivity.

**1900** Samuel Ruben is born in New Jersey.

**1903** Marie Curie is awarded the Nobel Prize in Physics for her contribution to the science of radiation.

**1908** Willard Libby is born in Grand Valley, Colorado.

**1912** Cosmic rays in the upper atmosphere are first discovered.

**1931** The Lawrence Berkeley National Laboratory is founded at the University of California, Berkeley.

**1937** The laboratory of Tree-Ring Research is founded in Tucson, Arizona.

**1939** Serge Korff notes that cosmic rays are continually generating neutrons in the atmosphere.

**1940** Martin Kamen and Samuel Ruben synthesize the carbon-14 isotope.

**1944** Marin Kamen has a fateful dinner with two Russian men, forever marking him as a potential Russian spy.

**1945** The Manhattan Project culminates with the atomic bombing of Hiroshima, Japan.

**1946** The first of the Dead Sea Scrolls are discovered in the Qumran Caves.

**1948** Martin Kamen is summoned to testify about Russian spying.

**1949** The results of Libby and Arnold's tests on radiocarbon dating are published in *Science*.

**1950** Willard Libby performs a radiocarbon test on a piece of linen from the Dead Sea Scrolls.

**1955** Hans Suess first reports the fossil fuel effect, which shows a slightly smaller amount of $^{14}C$ in and around large cities.

**1960** Willard Libby is awarded the Nobel Prize in Chemistry.

**1963** Atmospheric nuclear testing is banned.

**1965** The highest levels of atmospheric carbon-14 are detected.

**1967** The first calibration curve is published.

**1978** A committee is created in order to consider performing radiocarbon dating on the Shroud of Turin.

**1980** Willard Libby dies.

**1984** Scientists first report the ability to extract and sequence ancient DNA from the quagga.

**1988** Scientists at three labs date samples from the Shroud of Turin using radiocarbon dating methods.

**1991** The Dead Sea Scrolls are radiocarbon dated for the first time.

**1995** Further testing of the Dead Sea Scrolls is completed.

**2010** New radiocarbon dating studies confirm most theories about ancient Egyptian dates and events.

**2017** A groundbreaking report on cave art utilizes DNA testing in order to obtain accurate dates.

# Glossary

**accelerator mass spectrometry** A technique for measuring radiocarbon in samples by accelerating ions to very high speeds.

**archaeology** The study of history and prehistory, often through the analysis of physical items found during excavation.

**artifact** An object made by a human being, usually dating to a specific time period and used for historical or cultural study.

**atmosphere** The layers of gases surrounding Earth or another planet.

**calibration** The process of changing radiocarbon years to calendar years in the carbon dating process.

**calibration curve** A model used to analyze a radiocarbon sample against known dates, such as those determined through tree rings.

**carbon** An abundant, nonmetal element with the symbol C and the atomic number 6.

**carbon-12** The most common carbon isotope, against which the atomic mass of other elements is measured.

**carbon-14** A naturally occurring radioactive isotope of carbon, forming the basis for the radiocarbon dating method.

**contamination** Making something impure or corrupt by polluting, poisoning, or introducing unwanted contact.

**coral** A marine invertebrate mammal noted for its hard, stony exoskeleton that often forms within large reefs.

**cosmic rays** Atomic nuclei that hurtle through space, often striking Earth's upper atmosphere.

**cyclotron** An atomic particle accelerator machine developed by Ernest Lawrence in 1934.

**database** Any organized, efficient collection of data.

**Dead Sea Scrolls** A series of ancient Jewish manuscripts found within caves located near the Dead Sea, first discovered in the 1940s and 1950s.

**dendrochronology** The science or technique of dating events, environmental change, and archaeological artifacts by using the characteristic patterns of annual growth rings in timber and tree trunks.

**DNA** A type of organic compound (specifically, a nucleic acid) that contains the genetic information and is inherited, or passed from generation to generation, in all living organisms; the abbreviation for deoxyribonucleic acid.

**element** A substance that cannot be broken down any further using chemical means.

**epigraphy** The study of ancient epigraphs or inscriptions.

**fossil** The natural impression or remains of an ancient organism.

**geology** Scientific study concerned with Earth's structure as well as the rocks and other substances that make up that structure.

**half-life** The length of time it takes for half of a substance to decay.

**Manhattan Project** Research and development efforts undertaken by the United States government in order to produce the very first atomic weapons.

**organic material** Carbon-based compounds that originate naturally from plants and animals.

**organism** Any life-form, including an animal, plant, fungus, bacterium, or other living thing.

**paleography** The study of ancient handwriting and handwritten documents.

**periodic table** A grid-like arrangement of the chemical elements, organized by atomic number.

**radioactive decay** The process in which an unstable atomic nucleus emits radiation, causing it to lose energy.

**radiocarbon dating** A method to measure the radioactivity in an organic material in order to ultimately determine the age of that material.

**relative dating** The process by which one determines the relative order of past events, which can also help one determine the approximate age or age range of an object, like a rock or fossil.

**reservoir effect** A situation in which radiocarbon samples obtained from lakes, rivers, or other bodies of water may yield inaccurate dates.

**sarcophagus** A stone coffin, typically adorned with a sculpture or inscription and associated with the ancient civilizations of Egypt, Rome, and Greece.

**Shroud of Turin** A linen cloth, centuries old, that bears the image of a man said to be Jesus of Nazareth. It was radiocarbon dated in 1988.

**strata** Layers of soil or rock, formed at Earth's surface, that have consistent, unique characteristics.

**stratigraphy** A branch of geology that deals with the study of strata and the position of archaeological remains found within the rock layers.

**tree rings** Concentric rings that can be seen in the cross section of a tree trunk, each representing a year of growth in the tree's life.

# Further Information

## BOOKS

Conkling, Winifred. Radioactive! *How Irene Curie and Lisa Meitner Revolutionized Science and Changed the World.* Chapel Hill, NC: Algonquin Young Readers, 2018.

Huey, Lois Miner. *Children of the Past: Archaeology and the Lives of Kids*. Minneapolis, MN: Lerner Publishing, 2017.

Jackson, Tom. *The Elements Book: A Visual Encyclopedia of the Periodic Table.* New York: DK Publishing, 2017.

Mooney, Carla. *Chemistry: Investigate the Matter That Makes Up Your World.* White River Junction, VT: Nomad Press, 2016.

Weitzman, Elizabeth. *Mysteries of Stonehenge*. Minneapolis, MN: Lerner Publishing, 2017.

## WEBSITES

**Archaeological Institute of America: Ask the Experts**
https://www.archaeological.org/education/askexperts

On this page, you can ask volunteer researchers, professors, art historians, field archaeologists, and other experts any and all of the archaeology questions you may have.

**The Leon Levy Dead Sea Scrolls Digital Library**
https://www.deadseascrolls.org.il/home

This website offers high-resolution images of the scrolls that were discovered between 1946 and 1958. You can also learn about the scrolls and find more information about how the library was digitized and put online.

**Radiocarbon Laboratories**
http://www.radiocarbon.org/Info/lablist.html

Here you can find a complete list of all the current radiocarbon laboratories, including their contact information. Are there any near you? If so, you may want to get in touch with the lab and try to visit!

**SciStarter: Science We Can Do Together**
https://scistarter.com/

This is a database for "citizen science" projects, including collaborations between scientists and everyday people just like you. Here you can find projects in your area that are looking for participants. Sign up, contribute, and help real scientists come to real conclusions.

# Bibliography

Anderson, Lance. "Radiocarbon Dating Solves 48-Year-Old Museum Mystery in Peterborough." MyKawartha.com, February 16, 2018. https://www.mykawartha.com/news-story/8133076-radiocarbon-dating-solves-48-year-old-museum-mystery-in-peterborough/.

Balter, Michael. "New Dates for Egypt's Pharaohs." *Science*, June 17, 2010. http://www.sciencemag.org/news/2010/06/new-dates-egypts-pharaohs.

Callaway, Ewen. "Archaeology: Date with History." *Nature*, May 2, 2012. https://www.nature.com/news/archaeology-date-with-history-1.10573.

Chang, Kenneth. "Martin D. Kamen, 89, a Discoverer of Radioactive Carbon-14." *New York Times*, September 5, 2002. http://www.nytimes.com/2002/09/05/us/martin-d-kamen-89-a-discoverer-of-radioactive-carbon-14.html.

Curie, Mme. Sklodowska. "Radium and Radioactivity." In *Century* (January 1904): 461–466. College Park, MD: American Institute of Physics, 2000-2018. https://history.aip.org/exhibits/curie/article_text.htm.

Fessenden, Marissa. "Climate Change Might Break Carbon Dating." *Smithsonian*, July 27, 2015. https://www.smithsonianmag.com/smart-news/climate-change-might-break-carbon-dating-180956062/.

Hedman, Matthew. *The Age of Everything: How Science Explores the Past*. Chicago, IL: University of Chicago Press, 2007.

Jones, Nicola. "World's Largest Hoard of Carbon Dates Goes Global." *Nature*, July 11, 2017. https://www.nature.com/news/world-s-largest-hoard-of-carbon-dates-goes-global-1.22287.

Kanski, Alison. "Fossil Fuels May Bring Major Changes to Carbon Dating." Climate Central, July 28, 2015. http://www.climatecentral.org/news/fossil-fuels-major-changes-radiocarbon-19288.

Koppes, Steve. "UChicago Site of Radiocarbon Dating Discovery to Be Named Historic Landmark." UChicago News, October 5, 2016. https://news.uchicago.edu/article/2016/10/05/uchicago-site-radiocarbon-dating-discovery-named-historic-landmark.

Krakowka, Kathryn. "Science Notes—Updating Radiocarbon-Dating Techniques." *Current Archaeology*, February 5, 2018. https://www.archaeology.co.uk/articles/science-notes-updating-radiocarbon-dating-techniques.htm.

Libby, Willard. *Radiocarbon Dating*. Chicago, IL: University of Chicago Press, 1955.

Moskvitch, Katia. "Radiocarbon Dating Verifies Ancient Egypt's History." BBC, June 17, 2010. http://www.bbc.com/news/10345875.

Orgel, Hugh. "Age of Dead Sea Scrolls Determined Through New Carbon-Dating Method." Jewish Telegraphic Agency, April 4, 1991. https://www.jta.org/1991/04/04/archive/age-of-dead-sea-scrolls-determined-through-new-carbon-dating-method.

Panko, Ben. "Thanks to Fossil Fuels, Carbon Dating Is in Jeopardy. One Scientist May Have an Easy Fix." *Smithsonian*, December 7, 2016. https://www.smithsonianmag.com/science-nature/carbon-dating-crucial-scientific-technique-jeopardy-thanks-our-pollution-heres-easy-way-fix-it-180961345/.

Pedroncelli, Peter. "Scientists Use DNA Analysis and Radiocarbon Dating to Combat Ivory Trafficking." AFK Insider, February 15, 2018. https://afkinsider.com/141641/scientists-use-dna-analysis-radiocarbon-dating-combat-ivory-trafficking/.

Radford, Tim. "Emissions Threaten Age of Uncertainty for Carbon Dating." Climate News Network, July 24, 2015. https://climatenewsnetwork.net/emissions-threaten-age-of-uncertainty-for-carbon-dating/.

Renfrew, Colin. *Before Civilization: The Radiocarbon Revolution and Prehistoric Europe*. New York: Knopf, 1973.

Smith, Norman. *Millions and Billions of Years Ago: Dating Our Earth and Its Life*. New York: Franklin Watts, 1993.

Stromberg, Joseph. "A New Leap Forward for Radiocarbon Dating." *Smithsonian*, October 18, 2012. https://www.smithsonianmag.com/science-nature/a-new-leap-forward-for-radiocarbon-dating-81047335/.

Sullivan, Walter. "Serge Korff, Physicist, Dies at 83; Pioneer in Cosmic Ray Research." *New York Times*, December 2, 1989. https://www.nytimes.com/1989/12/02/obituaries/serge-korff-physicist-dies-at-83-pioneer-in-cosmic-ray-research.html.

Wade, Lizzie. "No Bones? No Problem: DNA Left in Cave Soils Can Reveal Ancient Human Occupants." *Science*, April 27, 2017. http://www.sciencemag.org/news/2017/04/no-bones-no-problem-dna-left-cave-soils-can-reveal-ancient-human-occupants.

Wheeler, Timothy B. "Long-Buried Oyster Shells Show How the Bay Sustained Early Americans." *Bay Journal*, March 8, 2018. https://www.bayjournal.com/blog/post/long_buried_oyster_shells_show_how_the_bay_sustained_early_americans.

Zimmer, Carl. "Neanderthals, the World's First Misunderstood Artists." *Independent*, March 7, 2018. http://www.independent.co.uk/news/science/archaeology/neanderthals-world-first-misunderstood-artists-cave-paintings-archaeology-a8230621.html.

# Index

Page numbers in **boldface** are illustrations.

# About the Author

**Kaitlyn Duling** believes in the power of words to change hearts, minds, and, ultimately, actions. A writer who grew up in Illinois, she now resides in Pittsburgh, Pennsylvania, where she has written over thirty books. She knows that knowledge of the past is the key to our future, and she wants to make sure that all students, teachers, and families have access to high-quality information.